For You

Andreas Seidl

Handover of Power

Global Version

Volume 16: Infrastructure

Imprint

Bibliographic information of the German National Library:
The German National Library lists this publication in the
German National Bibliography; detailed bibliographic data
are available on the Internet at http://dnb.dnb.de.

© 2022 Dipl. Pol. Theodor Andreas Seidl

Cover: Christiane Ebrecht
Translation: DeepL, Cologne
Production and publishing: BoD – Books on Demand,
Norderstedt

ISBN: 978-3-7562-1196-8

Acknowledgements

My thanks go to my family and friends who have made me who I am today. Special thanks to all those who supported me in writing this book. I would like to thank all my classmates, teachers, fellow students, lecturers, demonstrators, activists, colleagues, companies and countries with whom I have had the privilege of sharing the experiences from which all the ideas in this book have emerged. I would like to thank the staff of Books on Demand for their kind helpfulness. I thank the citizens of Seligenstadt for the harmony and solidarity in which I was able to write.

Foreword

This policy concept contains a variety of proposals for possible political reforms. It can be peacefully and democratically adapted to any current political system of any state in the world, but also to political systems in families, clubs, associations or companies. Wherever humans make or submit to rules that manage living together, the following proposals can be helpful. Readers who find the proposals so helpful that they would like to implement them together with like-minded people can contact the author. The contact form on the last page can be used for this purpose.

Faults and defects

I ask for your understanding that this volume was not professionally proofread. I could only afford professional proofreading for the summary. Spelling errors and unfortunate phrasing may therefore occur. As soon as this volume has sold enough to pay for a professional proofreading, it will be done. After that, a new edition will be published.

English version

Please understand that this volume has been translated automatically. I could only afford a professional translation for the summary. Poor wording and spelling errors may therefore occur. In case of doubt, the German version shall prevail. As soon as this volume has sold enough to pay for a professional translation, it will be done. After that, a new edition will be published. It was more important to me that no one in the world should have an information advantage than individual translation errors in the complete work.

References
If something has been quoted directly, it is set in italics. If the headings contain footnotes, the sources for direct and indirect quotations apply in the chapter for which the heading stands. Otherwise, quotations or source references are directly at the word or at the end of the sentence or paragraph. This book contains parts of text based on the Federal Constitution of the Swiss Confederation of 18 April 1999 (as of 12 February 2017), abbreviated to BV[1] and the Constitution of the Canton of Bern of 6 June 1993 (as of 11 March 2015), abbreviated to KV[2] .
If the constitutional paragraph, or individual paragraphs thereof, are based in whole or in part on extracts from the BV or KV, this is indicated in a footnote. The references to the corresponding footnotes for constitutional paragraphs are usually found after the heading of the affected chapter and sometimes in the body of the text. Articles used in the Swiss constitutions are listed in the footnote with a number after the title of the constitutional paragraph. Example: §123 Sample title: BV Art.123, KV Art.123.
All internet sources are fully cited in the footnotes. They were last accessed on 30.09.2021. All literature sources are also listed in full in the footnotes.
All references to tasks undertaken by other ministries and described in more detail there are given in footnotes. Example: Model Ministry - 1.2.3 Model Chapter.
All footnotes are to be viewed in comparison to the respective source, so-called indirect quotations. Direct quotations are set in italics, but hardly ever occur. The source reference is intended to enable further investigation and to take copyright

---

1 This is not an official publication. Only the publication by the Swiss Federal Chancellery is authoritative. https://www.fedlex.admin.ch/eli/cc/1999/404/de On 14.12.2021
2 This is not an official publication. The Bernese Official Collection of Laws is authoritative. https://www.belex.sites.be.ch/frontend/versions/2420?locale=de#ART71 On 16.12.2021

into account.

Table of contents

13

1 Goals of the Ministry of Infrastructure

The Ministry of Infrastructure pursues the goals of environmentally friendly and fast mobility, home ownership rates above 90% and import independent energy supply from renewable energy sources.

Every citizen living inland should be free to choose whether to live in the city or in the country. The place of work should not be the decisive factor. Therefore, it is necessary to drastically increase the travel speed for commuters. This will be possible through the technologies of the Maglev above ground or underground and the flying car in the air, as both means of transport have a travelling speed of about 500km/h. People's Innovation Company[1] and industrial communities will be established,[2] which will produce infrastructure construction machines on the one hand and means of transport using this infrastructure on the other. At first, this technology is developed and marketed domestically. This will turn the country into a permanent showcase for new innovative products that can be bought here. The aim is to use the profits from international marketing to replace taxpayers' money. The Ministry of Infrastructure is making its contribution with the innovative transport of goods and persons by land, sea and air, even into outer space. The second contribution comes from the Construction Team building the factories for People's Innovation Company.

In the long term, every nationals should own his or her own home so that the Unconditional Basic Income[3] does not have to be spent on rent. To this end, the Ministry of Infrastructure is gradually increasing the number of houses with the help of the housebuilding programme. The Ministry of Infrastructure uses the Real Estate Directory[4] to determine where these houses should be located. There, municipalities have the opportunity to expel residential building land. The citizens choose their locations. As soon as an economical and integrable number of houses is reached at a location, the Construction Team with

---

1 Ministry of Innovation - 10 People's Innovation Company
2 Ministry of Innovation - 10.4.1 Industrial communities
3 Ministry of Finance - 6 Unconditional Basic Income
4 Ministry of Digital - 12 Directories

infrastructurators arrives and builds the housing estate as if on an assembly line.

One goal of the Ministry of Infrastructure is to build industrial communities for the construction of infrastructure components and machinery. The components are used to prefabricate roads, tunnels, bridges, pipes and pipelines in individual parts according to a modular principle, deliver them and assemble them on site. Infrastructure machines are needed to assemble these puzzle pieces. These infrastructurators are a mobile multi-tool. All current construction machines can be placed on platforms here to have the right construction machines ready at any construction site at any time. There are infrastructurators for land, water and air.

The Ministry of Infrastructure's goal is to work towards making energy supply completely or almost free. On the one hand, this goal is achieved through energy self-sufficient houses. On the other hand, energy is generated for companies and conurbations from power plants that use only renewable energies and thus only incur maintenance costs. The goal is to use only renewable energies that do not cause any damage even when used over the long term.

2 Departments

The departments are divided into sub-departments and enumerations are usually considered as their individual units. Many tasks of some departments are completely taken over by other ministries as a service.

2.1 Central Department

Part of the Central Department is the Reception Office with the Courier and Mail Room, which directs all concerns, broadcasts and visitors to the appropriate place in the ministry.

2.1.1 Staff

The Human Resources Department is responsible for staff development and planning. For this purpose, it takes care of the recruitment of junior staff, intern and trainee programmes as well as the selection procedures for employees and special selection procedures for applicants with disabilities. For politicians and employees, the department prepares a job plan. In all its tasks, it works in voting with the personnel board.[5]

All other personnel matters are transferred to the relevant ministries. The Ministry of Education is responsible for the training and further education of employees for the state service.[6] The Ministry of Labour takes over the service law.[7] This includes the labour and collective bargaining law of the employees of the state service, remuneration, personnel administration of all careers and employees, flexitime, holiday and sick leave, working time with or without flexitime in part-time or full-time at the place of work or in home work. The Ministry of Finance's Pay Office takes care of employees' salary, expenses, travel and relocation costs.[8]

The Ministry of Education provides childcare for all employees in the state service.[9]

The Ministry of Health is responsible for the occupational health service.[10] It ensures occupational health management, deals with the treatment, education and prevention of occupational accidents, controls and provides occupational health and safety through the health auditors[11] of the Company Auditing Agency[12] .

---

5 Ministry of State Organisation - 2.1.1.1 Personnel board
6 Ministry of Education - 2.1.1.1 Education and training for the state service
7 Ministry of Labour - 4 State enterprises, 13 Labour Directory
8 Ministry of Finance - 2.1.1.1 Staff remuneration
9 Ministry of Education - 2.1.1.2 Childcare for employees in the state service
10 Ministry of Health - 2.1.1.1 Occupational Health Service
11 Ministry of Labour - 20.7.2 Health auditor
12 Ministry of Labor - 20 Company Auditing Agency

2.1.1.1 Housing assistance for state service employees

Housing assistance for state service employees includes the provision of housing within 10km of the service. State housing can be rented or purchased with a preferential right. This excludes Planned Economy housing. Other housing can be found through the Real Estate Directory. If there is no housing, a home can be built through the housebuilding programme and purchased on a hire-purchase basis.

2.1.2 Organisation

The ministries of media, security, justice, finance, labour, state organisation provide audit services for quality management in the ministry, evaluation of work performance, revenues and expenditures, as well as corruption prevention, sabotage protection and, if necessary, disciplinary matters.[13]
The Ministry of Labour regulates procurement law and ensures corruption-free state orders and procurement.[14] The Ministry of Finance organises the annual budget vote and ensures proper accounting in each ministry.[15] It regulates budget procedures, budget law, staff budgets, departmental budgets, costs and cash management, and assists ministries in budget planning for the budget vote. The language service for translating talks or texts is provided by the Ministry of Education.[16]
The Ministry of Digital Affairs supports the supply of Information Technology.[17] In voting with the Procurement Office of the Ministry of Labour, it takes care of the procurement, provision, maintenance and service of technical devices and software. Much of this is produced in-house to ensure data protection in information and communication technology. Information technology and digitalisation officers audit and advise the ministries. Digital appointment calendar

13 Ministries of Media, Security, Justice, Finance, State Organisation - 2.1.2.1 Audit services
14 Ministry of Labour - 6 Procurement Office
15 Ministry of Finance - 8 state revenues, 9 state expenditure
16 Ministry of Education - 2.1.3 Language Service
17 Ministry of Digital Affairs - 2.1.2.1.1 Supply of Information Technology

and documentation services are provided as well as a digital policy archive including a library.

2.2 Management Department

The Management Department is the minister's department. With his office team, he provides policy planning and analysis for his ministry and coordinates the relationship between the nation and the municipality through exchanges with his deputies in the municipalities. He initiates cooperation with other ministries or citizens in committees and is supported by the Ministry of State Organisation.

The Ministry of Media Affairs, through its media service, provides press and public relations for the ministry, moderates civil dialogue, trains or provides a spokesperson for the minister, writes speeches and texts on request, and ensures the implementation of conferences and events.[18]

The Ministry of Digital Affairs is responsible for digital management and thus provides departmental management. It automatically produces business statistics, staff surveys and the current state of research through statistics. It automatically forwards proposals to the affected or empowered state employees. In document management, it ensures digitalisation and that ministries share forms with each other.[19]

2.3 Homeland Department

The Homeland Department oversees and supports the cooperation of the other departments and agencies with other ministries. Cooperation is always necessary when infrastructure projects are cross-cutting, such as the use of natural resources, rural and urban development, environmental protection and homeland protection. The Homeland Department oversees the State Utilities and Municipal Utilities Company and coordinates their cross-cutting operations in the areas of construction, transport and energy. Recreational facilities are

---

18 Ministry of Media Affairs - 2.2.1.1 Media Service
19 Ministry of Digital Affairs - 2.1.2.1 Digital Service

planned in cooperation with the Ministry of Family Affairs. The construction and maintenance of recreational facilities is overseen by the Homeland Department. It operates the Real Estate Directory in cooperation with the Ministry of Digital Affairs and the Institute of Geosciences and Natural Resources in cooperation with the Ministry of Education. In voting with the Minister of Infrastructure, it formulates templates for legal rules on homeland and environmental protection.

2.4 Building Department

The Building Department supervises and supports the Building Offices, Building Yards and the Construction Team in the performance of their duties. It operates the Infrastructure Directory in cooperation with the Ministry of Digital Affairs. Construction projects are notified and audited via the Infrastructure Directory. The Building Department is the interface between the Minister of Infrastructure and the building authorities as well as with all other ministries that wish to implement a building project. It coordinates nationwide building management between the ministries and the Building Yards. It supervises industrial communities and the housebuilding programme and reports regularly to the Minister of Infrastructure. In the event of a disaster, it coordinates the deployment of the mobile cities. In voting with the Minister of Infrastructure, it ensures the formulation of templates for law building specifications.

2.5 Traffic Department

The Traffic Department oversees the Traffic Office and the construction and operation of the digital infrastructure, earthworks and traffic networks. It supports the Traffic Office by coordinating cooperation with the responsible State Utilities or Municipal Utilities Company, Building Offices and Building Yards or Construction Team. It operates the Transport Directory in cooperation with the Ministry of Digital Affairs.

In voting with the Ministry of Foreign Affairs, all continental and international transport matters are forwarded to the responsible agency and compliance with international or continental requirements by the authorities is supervised. It formulates templates for transport and traffic law as well as for environmental protection and digital mobility in voting with the Minister of Infrastructure.

2.6 Energy Department

The Energy Department ensures security of supply, economic efficiency and environmental compatibility in energy supply and the development of new sources of electricity. It oversees the operation of the state's decentralised and centralised energy facilities. It operates the Energy Directory in cooperation with the Ministry of Digital Affairs. In voting with the Ministry of Foreign Affairs, it refers all continental and international energy matters to the responsible agency and oversees the authorities' compliance with international or continental requirements. In voting with the Minister of Infrastructure, it formulates the law templates on the energy transition and on the consumption and supply of energy.

3 Tasks of the Ministry of Infrastructure

The Ministry of Infrastructure has the task of providing a homeland for the majority of the country's inhabitants. To this end, committees are held with the affected citizens. To protect the homeland and the environment, the exploitation of the country's raw materials, their recycling and the disposal of waste are regulated. The task of urban and rural development is considered to be fulfilled when sufficient agricultural land is available for the nourishment and recreation of the population and nature. At the same time, there must be sufficient land for development with buildings for living, working and moving around. The Ministry of Infrastructure ensures a home ownership rate of over 90% with its housebuilding programme, and easy admission to residential, commercial and recreational space with the Real Estate Directory. All state construction activities are taken care of by the Construction Team or the local Building Yard.

To fulfil its tasks, the Ministry of Infrastructure operates the State Utilities at the national level and the Municipal Utilities Company at the municipal level. Through their services, all state buildings are administered, all citizens are supported with drinking water and digital data, waste is disposed of properly, networks are operated, and energy is generated, distributed and stored.

The Ministry of Infrastructure's task is to ensure fast and environmentally friendly movement through the country. To this end, it facilitates individual transport and public transport for suitable and up-to-date means of transport. The Traffic Office ensures safe traffic.

In addition, it is the task of the Ministry of Infrastructure to ensure a secure, affordable and environmentally compatible energy supply. To this end, it operates the energy transition, the Energy Directory and its own renewable energy generation facilities.

4 Homeland[20]

The Ministry of Infrastructure is responsible for creating all the technical conditions to provide the population with the homeland they want. This includes organising territorial cohesion within the national borders and the municipalities. This organisation takes place in the territorial planning, which the Minister of Infrastructure agrees with the people and his deputies with the citizens of their municipality in a committee[21] and votes on. As a basis for the negotiations in the committee, the Institute for Geosciences and Natural Resources surveys the land on, above and below the earth's surface.

The law and the corresponding planning of spatial planning emerge from the results of regional planning. Spatial planning includes the development of urban and rural regions and the basic supply of infrastructure for the mobility of persons,

---

20§196,1,2 Spatial planning: BV Art. 75, KV Art.33, §197,1,2
Surveying: BV Art.75a, §220,3h Agriculture: BV Art.104
21 Ministry of State Organisation - 9.6 Committee

goods and data. Spatial planning law includes the requirement to settle the land in an orderly manner, which is why building permits are necessary, and to use the land sparingly and appropriately for building and quarrying projects.

Spatial planning is the basis for the building code, which obliges municipalities to ensure sufficient recreational space and agricultural land. The Ministry of Health determines which areas are sufficient for recreation in voting with the Ministry of Family Affairs. The Ministry of Labour determines which areas are sufficient for agriculture in voting with the ministries of economy. Otherwise, the municipalities are required to take into account the needs of their inhabitants, companies and the environment. First and foremost, the environment must not be damaged, secondly, the inhabitants should live happily and finally, companies should be given the best possible conditions.

4.1 Homeland protection[22]

The Ministry of Infrastructure takes care in its planning and construction activities, as well as in building permits, that landscapes, sites of local interest, historical sites, natural and cultural monuments can be preserved. Which objects are of such municipal importance is determined by the municipal population in a committee. The people may also determine such objects of nationwide significance in a committee. Objects with nationwide significance can be purchased or expropriated by the Ministry of Infrastructure. Expropriated properties are compensated at the current market value.

4.2 Environmental protection[23]

For nature conservation, areas are expelled that may not be built on or cultivated and may only be entered by humans for recreation and education. Animal and plant species should be

---

22§190,9 Environmental protection : KV Art.32, §191,1,2,3 Nature and homeland protection: BV Art. 78
23§190,9 Environmental protection: KV Art.32, §193,1,2,3 Forest: BV Art.77, §191,5 Nature and homeland protection: BV Art.78

able to develop freely there and their populations should be protected.

Environmental protection includes maintaining the protective, beneficial and welfare functions of forests and water bodies for future generations. In voting with the Ministry of Health, it is checked how a sufficient protective function is compatible with a useful and welfare function. The utility function is specified by the Ministry of Labour.[24] The welfare function is determined in voting with the Ministry of Family Affairs. In order to preserve the forest, in towns and cities, a forest reserve consisting of a green belt around the town and fresh air corridors through the town or city is tendered. In rural areas, the fewer trees there are, the more priority is given to expulsions as building land. More productive agriculture through permaculture[25] can free up land for forestation.

Waters are protected by sewers, sewage treatment plants and the prohibition of the discharge of any substance except drinking water. Anyone using standing waters and waterways must ensure that no pollution occurs as a result of their use. Products intended for use with water must be tested for safety by the Company Auditing Agency and may only be placed on the market after successful testing. For example, it is forbidden to sell sunscreen that pollutes water, or to sell clothing that, when washed, releases microfibres that cannot be filtered in sewage treatment plants. Paint or fuels that pollute waters are prohibited for watercraft.

4.3 Raw materials

Raw materials may only be mined after approval by the Ministry of Infrastructure and after a successful audit by the Company Auditing Agency. The licensing procedures are carried out within the legal framework of mining. Permits are only granted if sufficient raw materials are available for future generations and waste and overburden is disposed of in an environmentally sound manner. Any raw material mining above and below ground that is connected to overburden

---

24Ministry of Labour - 19.9 Forest, hunting and forestry policy
25Ministry of Labour - 19.8.7 Nature-based agriculture: permaculture

must establish a purpose for the overburden in cooperation with the Ministry of Infrastructure. Soil samples are taken to determine what the overburden can be used for. Overburden that can serve as building material is marketed or stored in an accessible manner. Other overburden is used for terrain formation, to create reservoirs or to fill in valleys over which roads run. In principle, underground cavities must be refilled and opencast mines must be renaturalised after their use.

For all finite mineral raw materials, a raw materials strategy is developed in a committee with the people. Quotas are set that are adjusted to the state of research on how quickly renewable substitutes can be developed. The price of finite raw materials must cover the costs of research and development of renewable or fully recyclable substitutes.

In order to be able to stock a sufficient number of different raw materials, resource efficiency is increased as constantly as possible. The fewer raw materials that lead to the fulfilment of the same benefit, the more likely a product is to be approved by a seal of approval[26] from the Company Auditing Agency. The fewer finite raw materials products contain, the lower the price premium for developing substitutes. Products that are fully recyclable are preferred in approval to comparable products that are not. Once a recyclable product has been approved, the approval for comparable products that are not recyclable is withdrawn within 5 years.

4.3.1 Institute for Geosciences and Natural Resources[27]

The Institute for Geosciences and Natural Resources continuously controls the stock of raw materials in the country and carries out measurements at undeveloped sites. It sends its results to the Company Auditing Agency for verification and to the Statistical Office for statistics[28] . The institute ensures measurement of how fast which finite raw materials are consumed and when the raw material is used up. To limit consumption, the Minister of Infrastructure is

---

26Ministry of Labour - 20.7.4.2 Seal of approval
27§197 Surveying: BV Art.75a
28Ministry of Digital Affairs - 6 Statistical Office

allowed to interfere with economic freedom and increase the price. Any price increase must be immediately invested in the development of substitutes.

In cooperation with the ministries of Innovation, Education and Planned Economy, biotechnologically producible and biodegradable substitutes for all finite raw materials are being developed.

The institute is responsible for surveying on land, at sea, in the air and in space. It forwards all collected data to the Ministry of Digital Affairs for processing into statistics. Based on this data, permits for construction and mining projects are granted or not.

4.3.2 Agricultural land

The Ministry of Infrastructure cooperates with farmers for nature-based agriculture[29] . The aim of the cooperation is for farmers to carry out horticulture and landscaping on state-owned land. To this end, shrubs and trees with edible fruits and nuts are planted and cared for along the edges of roads and paths, as well as edible plants in parks. This provides additional free food for the population. Foliage and prunings from state-owned land and agriculture are processed into heat, electricity and fertiliser in biogas plants. The resulting bioenergy and biomass are available primarily to the municipality, secondarily to the people and thirdly to foreigners. The Ministry of Infrastructure manages this bioeconomy, which provides growth through perennial crops and supports the population with food, electricity and heat.

4.4 Urban development

The Ministry of Infrastructure operates a decentralised urban development policy through the deputy infrastructure ministers in the municipalities. In committees with the citizens of the municipality, they determine where in the municipality which building culture is permitted or prohibited. This includes

---

29 Ministry of Labour - 19.8.7 Nature-based agriculture: permaculture

whether and how development plans are created, whether and how development is intermixed or segregated with properties for different income levels and companies, and whether and how urban redevelopment should take place with the joint cooperation of all residents. For example, a small town could decide to gradually renovate its half-timbered houses in the old town together and thus save on labour costs and purchase skilled labour and materials more cheaply in an alliance.

In the local Building Offices in the Town Hall, the land registry is operated digitally and the digital infrastructure for Smart Cities is laid when new construction or expansion of roads, paths and buildings takes place.

The Ministry of Infrastructure lays down town planning law and land registry law for the entire country, which the municipalities must adhere to. Urban planning law specifies where cities may be built and how far they may extend. The Land Registry Law stipulates how land boundaries are to be drawn and that owners of land must be known by name to the Building Office.

4.4.1 Living[30]

The Ministry of Infrastructure, through its Building Department, operates a state real estate and housing industry that promotes home ownership through lease-purchase procedures of its purpose-built buildings. This promotion is granted to individuals and housing cooperatives. For the purpose of housing assistance, the Construction Team may be requested by municipalities with low-income or elderly populations to build or rehabilitate housing in cooperation with the municipal population throughout the municipality.

30§25,4 Property guarantee: KV Art.24

4.4.2 Home ownership rate[31]

The Ministry of Infrastructure aims for a home ownership rate of 90% of all nationals in the country. Social Villagers and residents of Barter Economy Zones[32] are exempt from this. Until the quota is reached, the housebuilding programme is operated and the proportion of second homes is limited. The proportion of second homes in a municipality may not exceed 20%, of the total housing units and floor area used as living space.
All Building Offices monitor the municipal home ownership quota in their municipality. If it is met, the housebuilding programme is terminated or not applied in that municipality and second homes are unlimited. The National Building Office monitors the national home ownership rate and coordinates the use of the housebuilding programme with the municipalities.

4.5 Real Estate Directory

In the Real Estate Directory, all flats, houses, offices, halls and factories in the country are given a profile. This digitises the land register system. Each entry in the land register corresponds to an entry in the profile in the Real Estate Directory. The graphics correspond to those of google earth 3D in the exterior view. The interior view of the buildings is simulated with the help of the data entered in the building permits on the basis of the building plans. If properties are to be sold or rented out, sellers or landlords can lend out an indoor virtualiser[33] at the town hall, which captures the interior with a 360° 3D camera. Afterwards, the users bring the device back to the town hall, where the data is read out and posted on the intranet.
Properties for sale or rent are shown in green, rented properties in yellow and sold properties in red. Anyone who wants to build a house can right-click on the selected area of the map and specify "build here". Then the potential builder can upload his building project. In the building project, virtual buildings are created using a 3D programme. You have to be

---

31 §198 Home ownership: KV Art.40, BV Art. 75b
32 Ministry of Barter Economy - 6 Barter Economy Zone
33 Ministry of Digital Affairs - 14.7 Indoor virtualiser

as precise as the building laws require. Anyone who forgets information cannot upload the building project at all. The programme then lists all missing information.

Newly uploaded building projects are automatically displayed and checked by the responsible Building Office in the Town Hall. New building projects that are uploaded and checked are displayed in blue and an automatic news item about the building project is sent to all residents within a radius of 50 metres. 30% of affected citizens can convene a committee through a veto quorum[34] . 80% of the affected citizens of a city or 60% of the people can prevent or change the construction of any property by voting.

Citizens can form groups to prevent or promote building projects. If they want to push ahead with building projects, they can work together in a group to raise the necessary funds and be joint builders.

Properties marked in green can be viewed from the inside by left-clicking on them and by right-clicking on them the contact details for landlord, seller or agent can be viewed. Every landlord or seller must conduct this business via the Real Estate Directory or at least indicate the rental price or sales price so that the business tax can be correctly accounted for. Tenants can also view a rental price history or purchase price history here. By rubber banding an entire area can be marked to retrieve the average or median price per square metre to buy or rent. In the case of the median, the complete table of the selected flats or houses in the marked area can be called up. Landlords and sellers have different rights and obligations in the different economic forms.[35]

Anyone can report themselves as looking for a home in the Real Estate Directory and submit a motions for a home. Anyone looking for a home can save their search settings and receive new listings on an ongoing basis. Those who want to participate in the housebuilding programme will receive a notification when one of their desired locations has been marked as a building site for the housebuilding programme.

---

34 Ministry of State Organisation - 9.5.14 Veto quorum
35 Ministries for Economic Affairs - Real estate sector

4.5.1 Real Estate Finder

Citizens or companies can use the Real Estate Finder to search for suitable properties. In their search, they can enter specific details or open a range that would satisfy them equally. Users can select keywords to describe the location or type of construction, specify the square metre and storey height or even search within a postcode area. Users have a field to allow or deny their data to be used to contact persons of a similar search. This is intended to make it easier to set up building communities or Residential Communities.

4.5.2 Residential Community Finder

Those who are relocating can use the Residential Community Finder to locate or create residential communities. Users have a choice for residential buildings and divisible interests. For residential buildings, users can select whether the building should be a single-family house, apartment building, high-rise building, farmhouse or a housing estate consisting of single-person houses. For each building, the child support or rental costs are indicated.

Interests can include current living situation, such as disabled, vegan, religious, unmated person, with child or retired. Other interests such as profession or leisure activities describe which employment opportunities can be shared, such as sports, making music or DIY. Via data retrieval consent, all entries from the user's profiles in all directories can be retrieved via an algorithm to find persons who have a similar or certain lifestyle and interests.

Users of the Residential Community Finder can form residential groups and build, buy or rent land or buildings together. If the group is large enough, an application can be made for the housebuilding programme. The lower the assets of all members of the residential group, the higher priority they will receive. Users can create their own construction and housing projects in a housing group.

As a standardised housing project, the Residential Community Finder offers old farms that can be developed into a residential community for self-sufficiency and can be purchased through a hire-purchase procedure. Suitable buildings are proposed and adapted to the number of people. The size and number of buildings on the farm determine how many persons can be supported and how many persons are needed to finance and run the farm.

4.5.3 Municipal housing exchange

Cities can use the Real Estate Directory to promote or specifically mix residents with similar living situations and interests in a neighbourhood. The city's municipality decides whether the urban housing exchange can be used voluntarily or is mandatory. Users can search for similar interests and abilities of their neighbours via data matching. To do this, they have to fill out a registration form at the beginning of using the housing exchange, which shows what one likes, what one can do, what one would never want to do without and what one finds obnoxious. For example, a user can cook and clean, likes handicrafts and DIY, cannot do without football and finds punk rock obnoxious. The urban housing exchange is the milder version of a cultural protection area and is intended to promote neighbourly friendships and avoid disputes.

4.6 Leisure

The Ministry of Infrastructure builds and operates facilities for the physical training and creative development of the population. Construction and operation is done in voting with the Ministry of Family Affairs.[36] The land owned by the state should, as far as possible, always also provide recreational benefits where sufficient citizens live.

Leisure centres adapted to the locality are established in all large cities. Large cities along rivers will have bathing lakes that are connected to the rivers via channels. Parking areas will be equipped with barbecue areas including connections for electricity and water as well as a fireplace and trim trails.

36Ministry of Family Affairs - 9 Leisure

Industrial sites that cannot be sold for more than a year are converted into multi-purpose areas and made available to the population until they can be sold. The local municipality is given the right of first refusal to create a Leisure Centre. For example, tarred roads become race tracks, concrete surfaces become fairgrounds, asphalt surfaces become an airfield, halls become event spaces and manufacturing halls become workshops. These centres are built by prisoners and financed by tax money. In all forests near housing estates with children, forest areas are expelled where children and youths are allowed to build tree houses and huts.

The Ministry of Infrastructure allows citizens to use public workshops and recreational areas, hold street festivals and visit the People's Restaurant.

4.6.1 Public workshops

Citizens can combine their skills and learn new techniques from each other free of charge and without obligation in the public workshops. These workshops are located in town halls, art or workshop rooms of educational institutions or similar public buildings with the necessary equipment. Opening hours are whenever the building is unused. Use must be registered with the building management or through the Real Estate Directory so that a person is given responsibility and keys. Public workshops obtain their materials from user donations and bulky waste. These workshops are accessible to children and adults alike.

4.6.2 Roofing

Citizens can lend out free of charge unused tent roofs from the People's Motor Vehicle[37] for their recreational events or have canopies built by the Construction Team. These canopies must be paid for by the citizens who want them. People's Motor Vehicle canopies are in a trailer and must be picked up by a person who has a Truck and a licence to do so. If the

---

37 Ministry of Media - 7.1.1 People's Motor Vehicle

citizens cannot find anyone, they can buy the service from the Social Service.[38]

4.6.3 Street parties

A street party may be held in any street at least once a year. For this purpose, a street closure for up to 24 hours can be requested from the Traffic Office. All residents within a radius of 100 metres immediately receive a message via their news service in the Persons Directory with the date and location of the street party. Residents can veto this, but only if the festival takes place more than once a year. No noise pollution or nocturnal disturbance of the peace applies on the night following the day of the street festival. The deputy infrastructure minister of the affected municipality can influence the licensing requirements at the Traffic Office. For example, several street festivals can be held at the same time to reduce noise pollution.

4.6.4 Recreational areas

Unused roads or areas owned by the state are made available to citizens free of charge. Road traffic licensing regulations do not apply there, which is why all kinds of means of transport or things can be operated there, even without approval by the Company Auditing Agency. All those affected must be warned of dangers and asked for permission. All actions are at your own responsibility. If the municipality detects offences or misdemeanours on recreational areas, a vending machine can be set up where users have to register with their identity card. In addition, a video surveillance system can be set up and connected to the nearest police station.

4.6.5 People's Restaurant

Municipalities can commission a People's Restaurant from the Ministry of Infrastructure. All bars and restaurants in the city can open a branch there, which costs 20% of the profits as rent.

---

38 Ministry of Planned Economy - 9.4.1 Social Service

If no profits are made, the rent is waived. Like in a department store with several shops, the restaurants share tables, chairs and a dance floor including a stage on one floor in the middle. On the sides, each restaurant has its own bar and counter with bar stools. Food is delivered from the participating restaurants. Snacks can be prepared on site.

Restaurants that host regulars' tables and have a branch in the People's Restaurant organise regulars' table rounds in the People's Restaurant in voting with the other branch owners. The aim is to bring regulars' tables into a larger circle so that they do not develop such radical manners and opinions, but have the chance to hear other opinions so that a compromise can emerge. In this way, even city dwellers practise democratic discourse in their free time, in which parties can participate. To this end, there are guest speakers from parties and petitions or initiatives.

The homeless are fed there on religious and state holidays in order to live charity as a value to which the people are committed. All restaurants collect food donations from the population for this purpose.

4.7 Building management

All state buildings and the rooms in them are built, renovated and administered by the Ministry of Infrastructure. The premises are to serve the people around the clock as far as possible. The building administration is located at the Building Yard and is responsible for ensuring that all premises are rarely empty. First and foremost, state tasks must be fulfilled; secondly, citizens may use the premises. Thirdly, companies of the Planned Economy, Social Market Economy, Barter Economy and finally the Free Market Economy.

4.8 State Utilities

The State Utilities are operated by the Ministry of Infrastructure. They ensure the cooperation of the Municipal Utilities Companies in the alliance as well as the operation of the nationwide networks and the facilities for nationwide energy supply. They operate the meteorological service and forward its data to transport companies, means of transport, energy suppliers and farmers.

4.9 Municipal Utilities Company[39]

Municipal Utilities Companies are the municipal institutions of the Ministry of Infrastructure. They are responsible for the decentralised supply of the municipality with data, electricity, heat, water and the disposal of liquids, solids and gases. All Municipal Utilities Companies are part of the national alliance and are supported in their supply by the national networks, waste management companies, electricity generators and data storage facilities. Municipal Utilities Companies are the responsible point of contact for citizens, whether services are provided on a municipal or national basis.

Municipal Utilities Company, in cooperation with the Ministry of Digital Affairs and the People's Innovation Company Intranet, operate the facilities for wired and wireless transmission of digital data of the Intranet.

In cooperation with all citizens, the Municipal Utilities Company operates facilities for energy production, such as solar panels on roofs and biogas plants for faeces and organic waste. Wind power and hydroelectric plants enable the Municipal Utilities Company to generate additional electricity. They store energy decentrally in the form of biogas, zeolite, compressed air, hydrogen and water storage. Biogas, zeolite and hydrogen can be used to generate heat or electricity.

Municipal Utilities Companies have the sole right to authorise wells for the production of drinking water. They can prohibit the operation of any privately or company-owned wells to produce drinking water if the municipality's drinking water supply is at risk.

Disposal plants are operated by the Municipal Utilities Company to process waste in waste recycling plants or waste water in sewage treatment plants. Persons who dispose of waste gases through chimneys must fit suitable filter systems

39§192,2,4 Water: BV Art. 76, KV Art.35, §199,3 State enterprises

to make the waste gases harmless to the environment. The Municipal Utilities Company supervises the installation of suitable filter systems for private persons and in cooperation with the Company Auditing Agency for companies.

In any national administration and management by the Ministry of Infrastructure, attempts are made to have the Municipal Utilities Companies cover their costs and act independently in their personnel policy. Since the use of the grid is mostly necessary, the Municipal Utilities Company has to pass on the grid fees to their customers. The Ministry of Infrastructure sets the cost-covering prices for the networks and adds 10% as profits. However, if cities manage to support themselves without the national networks, they are able to save these costs. This would be at the expense of state profits, but self-sufficiency by the residents has priority. Municipal Utilities Companies are owned by the citizens of a municipality. They can use them as general property to support themselves independently.

Among other things, the Municipal Utilities Company also operates the public pay and display parking areas. For this purpose, there is a special parking payment system for cities. Parking spaces are monitored with cameras that record licence plates of parked cars and automatically debit the parking fees via the tax account[40] at People's Bank .[41]

4.9.1 Water supply[42]

The Municipal Utilities Company protects the water resources by ensuring that no contaminants can enter the groundwater. Regular samples of drinking water and treated wastewater are taken by Municipal Utilities Company staff and tested for safety in their own laboratories. Samples are sent to the Company Auditing Agency's health auditors at regular intervals to be newly and more extensively tested. The Company Auditing Agency inspects the drinking water

---

40 Ministry of Finance - 5.5 Tax account, 11.6.2 Tax account for companies
41 Ministry of Finance - 11 People's Bank
42 §192,1,4,5,6 Water: BV Art. 76, KV Art.35

production and wastewater treatment plants annually and also unannounced at irregular intervals.

The Municipal Utilities Company sell drinking water and service water at cost price plus 10% profits. They raise prices when water shortages are imminent and increase the price increase at regular intervals according to the circumstances until the water shortage has been eliminated through more economical water consumption. In addition to the price, private individuals and companies are given advice on how to save water. Companies can be obliged to save water in the event of a water shortage and must have their implementation plans audited by the Company Auditing Agency within 3 months.

The water supply is guaranteed by the Municipal Utilities Company and can additionally be supplemented by wells owned by private individuals and companies. The use of the wells can be restricted in whole or in part by the Municipal Utilities Company if this would inevitably lead to water shortages in the future.

Municipal Utilities Companies are part of a nationwide alliance and release water to each other in case of water shortages and can be forced to do so by the Minister of Infrastructure if necessary. All Municipal Utilities Companies regularly report their water reserve levels and their renewability to the Ministry of Infrastructure, where the data is processed. An algorithm evaluates the data and proposes a distribution. This distribution enables sufficient renewal capacity of the drinking water reserves and a similar drinking water price throughout the country. Price differences are only intended to reduce water consumption in affected municipalities. Municipalities with less water receive water from municipalities with more water.

Foreigner companies are not allowed to extract drinking water, and springs and wells may not be sold to them. The use of fresh water can be prohibited to companies in case of imminent water shortage. The right to use international water resources is agreed by the Ministry of Infrastructure in voting with the Ministry of Foreign Affairs and affected states and submitted to the people for a vote.

4.9.2 Waste disposal[43]

The Municipal Utilities Company operates waste collection and the State Utilities operate waste recycling plants. Waste that can be processed well together is picked up jointly by the waste collection service. In the short term, refuse bins, refuse sacks and containers for old clothes, old glass and batteries are used for this purpose. In the medium term, household waste is delivered directly to the wastewater treatment plants via the waste disposal network and collected there by the waste collection service. Other waste is delivered to recycling centres run by the Municipal Utilities Company or disposed of via bulky waste.

A national cycle of recyclable materials is to be created through waste recycling. All recyclable materials that are consumed in the country should also be reused here if possible. In order to be able to recycle waste as easily as possible and to separate waste completely, companies must adapt their products and packaging to the requirements for easy recycling in order to obtain the necessary test seal .[44]

4.9.2.1 Pricing in the cost of pollution

All products that generate waste pay a levy to the Ministry of Infrastructure that covers 110% of the disposal costs. For example, the producer of a wrapped candy bar must pay a price for the plastic wrapper so that the plastic can be recycled back into petroleum or equivalent plastic. A price must be paid for the chocolate bar that is sufficient to clean the wastewater from the resulting faeces and toilet paper. The remaining 10% goes to the Ministry of Infrastructure as profits. This means that any waste disposal costs are already priced into the sale price. This means no one has to pay for rubbish collection. This is already done by the producers of any goods that become waste in whole or in part or produce packaging waste. Likewise, the price includes any pollution costs that are incurred during production, use and disposal. The cost of pollution covers all

43§190.5-8 Environmental protection: KV Art.36, §199.3 State enterprises
44Ministry of Labour - 20.7.4.2 Seal of approval

costs that must be incurred to fully undo the pollution. If this is not possible, the pollution is banned.

4.9.2.2 Waste recycling plants

In waste-to-energy plants, waste is incinerated, composted or processed to be reused. Waste incineration is used for district heating and electricity generation. Biogas plants generate electricity or feed the local gas grid. Recyclables, such as metal, glass, paper and plastic, are processed and reused. Toxic substances are rendered harmless or stored in special storage facilities until their rendering harmless has been researched.

Waste that is harmful in the long term must be avoided. If it still accumulates, the producers have to pay for the costly methods to render the waste harmless or pay for harmless storage and research to render it harmless.

The sewage systems of a city end in wastewater treatment plants. Here, all solids are screened out and sent for waste recycling. Residual wastewater is treated by microorganisms, gases and UV light until it is harmless and discharged into the local river system.

Huge sewage treatment plants are being built at all river mouths into the ocean to purify the entire river before it flows into the sea. These river treatment plants are offered as export goods.[45]

4.9.2.3 Bulky waste

What counts as bulky waste is determined by the Minister of Infrastructure in voting with the people. His deputies in the municipalities can also vote on their own requirements with the citizens of the municipality.

Citizens can order bulky waste from the Municipal Utilities Company once a year free of charge and must pay for each additional pickup. The cost-covering price plus 10% profit applies. During the pickups, citizens indicate what is to be collected. The Social Service of Planned Economy collects

45 Ministry of Foreign Affairs - 7.3.3.5 International water supply

usable or repairable items that are not already in stock and recycles them.[46]

There is also a free bulky waste collection on 2 days a year, where all citizens can place their bulky waste at the roadside the day before. As soon as bulky waste is placed at the roadside on this day, it is general property and anyone may help themselves. The first person to touch an item owns it. This bulky waste collection is intended to promote the reuse of used items and reduce the cost of living for the population. Public workshops can use this opportunity to fill their warehouses and the Social Villages to advance their luxury supply. Remaining bulky waste is collected by the waste collection service.

4.9.2.4 Waste collection

The Municipal Utilities Company organises annual weekend waste collection campaigns in every city. The frequency increases as pollution levels rise. Primary and comprehensive schools are required to participate in the events and citizens can take part on an honorary service to rid their environment of litter lying around. For this purpose, there is a litter-marking function in the People's Navigator[47] , which every citizen should use if he sees so much litter in a place that he could not clean it up unceremoniously on his own. State employees, especially those on patrol, are required to regularly mark places where rubbish has been found, which are then cleaned up during the rubbish collection campaigns.

5 Building[48]

The Ministry of Infrastructure is responsible for the construction and maintenance of state buildings, such as networks, transport routes, water supply, waste processing and energy production or storage facilities, squares and buildings. The National Construction Team is responsible

---

46 Ministry of Planned Economy - 5.5.1 Recycling of goods, 10.5.7.3 Upgrading company

47 Ministry of Digital Affairs - 11.4 People's Navigator

48 §199.3 State enterprises, §205.4 Energy policy: BV Art. 89, §206.3 Transport of energy, data, water and waste water, §226.6 Housing and

for construction and major renovation work. The municipal Building Yard administers the buildings and infrastructure in the area. Its employees do minor repair work and janitorial services. Other ministries are required to have their construction projects handled by the Construction Team and building management by the Building Yard. The Building Department in the Ministry of Infrastructure is the central point of contact for all building projects of other ministries. The Building Offices in the town halls are the contact points for all building projects of private individuals and companies in the respective municipality where the building project is to be implemented.

Through its own enterprises and industrial communities, the Ministry of Infrastructure ensures that private housing and state buildings are built more cheaply and more quickly. For this purpose, construction modules are pre-produced on an assembly line and delivered to the construction site, and construction machines are manufactured that combine all the features of construction vehicles at one construction site. The housebuilding programme uses these construction methods on a large scale until the home ownership rate reaches 90%.

The Ministry of Infrastructure sends architects of the construction team abroad to inspect and supervise foreign construction matters of buildings of the state abroad.

5.1 Building Office[49]

The Ministry of Infrastructure operates a National Building Office and many municipal Building Offices. The National Building Office is responsible for building projects that span municipalities, coordinates the municipal building offices in their cooperation and ensures uniform requirements throughout the country. The nationwide uniform requirements are regulations on which land areas may be built on and on construction rationalisation. The National Building Office is the central office for building rationalisation. There, standards are set for the entire life cycle of a building. The standards range

home ownership promotion: KV Art.40
49§226.3 Housing and home ownership promotion: BV Art. 108

from planning to realisation to management and establish a uniform language, digitalisation and system. Planners, builders, companies and providers can thus collaborate more easily and digitally with each other and the Building Office. The national requirements can be extended by the municipal Building Offices and adapted to the circumstances of their municipality. The Building Offices in the municipalities are responsible for all building projects that are implemented within the municipality.

5.2 Building Yard

The Building Yard is the representative of the construction team in each municipality. It stores and maintains machinery and materials needed by the Building Yard and construction crew. In cooperation with the Building Yard, citizens can carry out building projects in their municipality. In cooperation with the Construction Team, individual employees of all Building Yards can reinforce the Construction Team's construction workers, or the Construction Team temporarily reinforces the forces of a Building Yard until a construction project is completed. Each Building Yard maintains a caretaker service that looks after all state buildings in the vicinity and notifies the Construction Team in the event of major construction work.

5.3 Building specifications

The Ministry of Infrastructure sets requirements for building matters inland through the Building Law and the Building Code. Building law stipulates that structures must have a structural design that does not fail immediately even in the event of an earthquake, storm or flood, but provides protection for the occupants or at least withstands it until the occupants can get to safety. The building and systems engineering must ensure that the use is not dangerous and also does not put third parties at risk. For example, poorly laid cables for data, electricity, gas or water could cause damage.

To this end, building engineering is subject to regular training and inspection by Company Auditing Agency technical auditors.[50] Buildings are to be inspected at various specified stages of construction by a chartered architect, civil engineer or technical auditor. The building code also stipulates sustainable construction, which requires building components to be as easily replaceable, extendable and reusable as possible.

The building code stipulates that different buildings are subject to different building codes and must meet certain norms to do so. For example, houses made of clay, wood, reinforced concrete, bricks, building blocks made of hemp and lime or composite materials are built according to different types and are subject to different norms, without which they could lose their statics or building safety. These norms are laid down in the building-related building law. In the future, building materials must no longer release CO_2 during production and must be recyclable. Buildings must be energy-efficient. Solar cell modules may only be erected on building roofs, not on fields. Large wind turbines that produce cast shadows and noise may only be erected off the coasts.

For the construction of state buildings, the procurement of building materials via the Procurement Office[51], the completion of construction work by the Construction Team applies. Only building blocks made of hemp and lime may be used wherever possible. In exceptional cases where the Procurement Office or the Construction Team are not sufficient, contracts may be awarded to companies in compliance with the Public Procurement and Contract Regulations for Construction Works .[52]

The state construction industry through the Building Yards and construction crews works in research association with colleges to discover basic principles and new developments in construction research. The construction industry of the Planned Economy is obliged to participate, for construction companies of other economic forms it is voluntary. However, the Minister of Infrastructure can extend it to these companies

---

50 Ministry of Labour - 20.7.4 Technical auditor
51 Ministry of Labour - 6 Procurement Office
52 https://www.fib-bund.de/Inhalt/Vergabe/VOB/

by law.

Compliance with the requirements is checked by the Building Offices before building permits are issued. They apply equally to state, private and corporate buildings. The requirements mainly concern energy efficiency, urban regeneration, networks and transport routes.

5.3.1 Energy efficiency

The Building Offices are responsible for measuring the energy consumption of households, commerce and industry in their municipality and for informing and advising property owners about energy saving opportunities. To prevent environmental damage, the Building Office can set a deadline for owners by which renovation measures must be completed. The innovation auditors of the Company Auditing Agency[53] suggest to property owners which products and measures are suitable and explain to them the possibilities for construction and financing through environmental innovations.[54] In the case of new buildings, the requirements must be met from the outset in order to obtain planning permission. The innovation auditors also advise building owners on suitable products and measures.

The requirements are to insulate buildings in such a way that they emit as little heat as possible to the outside. On the one hand, this saves heating costs and, on the other, it does not heat up the environment. The insulation should also let in as little ambient heat as possible, which reduces cooling costs and heats up the environment less. The insulation is provided by walls that are built from bricks consisting of hemp and lime. This makes it possible to use a building block that binds more CO_2 than it consumes and grows back quickly.[55]

New buildings should be built in such a way that heat exchangers are operated via water tanks in the ground. In the basement is the energy centre for electricity and heat. Zeolite[56]

---

53 Ministry of Labour - 20.7.5 Innovation auditor
54 Ministry of Innovation - 9.10.3 Environmental innovations
55 https://www.hanfstein.eu/
56 https://www.energie-experten.org/heizung/heizungstechnik/pufferspeicher/zeolith-waermespeicher

, liquid salt[57] and a heat exchanger that can generate electricity from heat serve as energy storage. These technologies have low technical and digital requirements and can therefore be used in private buildings. As radiators and hot water boilers, the Ministry of Digital Affairs offers modules that perform computing and give the resulting waste heat to consumers free of charge.[58]

Zeolite modules on the southern façade are dried there by the sun in summer to release heat in winter in the heating system by adding water. Devices for hanging plant modules are attached to the other parts of the façade. They provide additional insulation, improvement of air quality, noise protection, biomass for biogas production, food in the case of useful plants and amusement in the case of ornamental plants. Solar panels are installed on the roofs to generate electricity and hot water. In this way, buildings can support themselves with sufficient electricity and heat through renewable energies from their immediate surroundings, if possible.

Exceptions apply to listed buildings.

5.3.2 Urban redevelopment

The municipal Building Offices are also responsible for energy-efficient urban rehabilitation. They ensure the elaboration of urban building projects that guarantee mobility for persons that is as barrier-free as possible without obstructing the area with obstacles such as roads and rails. Pavements of streets, paths and squares are to be replaced by suitable solar cells wherever possible, street lamps are to generate their own electricity via solar cells and store it by means of batteries in the pole. Building Offices are obliged to negotiate and coordinate urban construction projects with the affected population through the Deputy Minister of Infrastructure in a committee.

57https://www.solarify.eu/2020/01/29/857-energie-kuenftiger-heizsysteme-in-salz-speichern/
58https://qalway.com/en/ecological-heat/

5.3.3 Networks and transport routes

The National Building Office is responsible for the planning and approval of networks and transport routes. The Construction Team is responsible for the construction and upgrading as well as all engineering structures, such as bridges, tunnels, noise barriers and environmental protection measures. The Construction Team administers its material resources and personnel in cooperation with the Building Yards of the municipalities.

Committees and voting are held for the planning of roads and other routes above ground. In the case of construction projects with nationwide implications, the people are involved; in the case of municipal construction projects, only the population of the municipality is involved.

5.4 Building permits[59]

Building permits are issued by the National Building Office for the development of nationwide networks, transport routes and for the construction of new towns, Social Villages, Asylum Villages and Barter Economy Zones. The municipal Building Offices issue building permits for constructions within their municipality. In case of doubt, the National Building Office decides which municipal Building Office is responsible for building projects within the boundaries of several municipalities.

To obtain a building permit, builders contact the responsible Building Office and submit an application describing their building project. As soon as the Building Office receives the application, the building project is published. On the one hand, the publication takes place at the location of the building project in poster form. On the other hand, submitted building applications are also published immediately in the Real Estate Directory and affected residents are notified.

Building permits may not be granted in endangered areas. Endangered areas are flood zones, areas of departure for

---

59§186.2 Peaceful separation, §226.3 Housing and home ownership promotion: BV Art. 108

avalanches or mudflows or volcanically active zones. Building projects must be suitable for the subsoil.

Depending on the economic form, further building requirements may apply that must be observed. The environmental compatibility and financial security of the building project are always checked.

Private or corporate building projects are approved if they comply with the building law and building regulations and the surrounding population does not meet a veto quorum while the building application is being examined.

State building projects are always put to a vote and can be negotiated in a committee if they are controversial or have been rejected once. Once the environmental impact and buildability tests have been passed, the building application is put to a vote. All affected citizens then have one voting week to cast their vote. If the citizens vote in favour, the building permit is granted.

5.5 Building inspection

Building inspection starts with the examination of the building applications in the Building Offices, which is completed with the building permit. After that, the Company Auditing Agency's technical auditors inspect the building as soon as it is completed. Any defects found are listed in the inspection report and must be rectified. Once rectified, a re-inspection is carried out. Only when structures have been inspected without defects do they receive the operating permit with the inspection report. A copy of the data is sent to the responsible Building Office and stored in the profile of the building in the Real Estate Directory. Structures are inspected every 20 years by the technical auditors and the innovation auditors.[60]

---

[60] Ministry of Labour - 20.7.4 Technical auditor, 20.7.5 Innovation auditor

5.6 Needs assessment

Every citizen can report a need, be it a pothole in front of the house or a highway that would make it easier to get to work. Anything is possible on the virtual map of the People's Navigator[61] . However, it is important that as many citizens as possible express the same wish. Only one vote would be needed for the pothole, because it endangers road safety. For the highway, on the other hand, several million. Some proposals are necessary to ensure road safety and will be remedied as soon as possible. Construction projects such as a highway or bypass are examined by the responsible Building Office if 30% of the affected citizens vote in favour and are included in the annual budget vote[62] together with a time and cost calculation. If 65% of the citizens vote in favour of the state construction project, it is built.

5.6.1 Citizens' building project

Citizens can submit initiatives for building projects to the Building Office, just as they can for laws.[63] Once the quorum is reached, the building project is examined by the responsible Building Office and approved or not by voting. Building projects of the people are handled by the national Building Office; building projects of citizens of a municipality are handled by the municipal Building Office. Before voting, those entitled to vote clarify the financing. Either donations can be collected, savings from the Ministry of Infrastructure can be used or tax funds can be used. The use of tax funds can only be decided in the budget vote. The Minister of Infrastructure decides on the use of savings.

---

61 Ministry of Digital Affairs - 11.4 People's Navigator
62 Ministry of Finance - 9.5 Budget vote
63 Ministry of State Organisation - 9.10.11.7 Citizens' Initiative

5.7 Infrastructure Directory

The Infrastructure Directory is the digital platform for constructing, renovating or refurbishing buildings from the Real Estate Directory. Each construction project is given a profile, the content of which is transferred to the corresponding profile in the Real Estate Directory once the building is completed. In the Infrastructure Directory, the building project is presented as a digital three-dimensional animation that can be built with the help of construction workers from the building groups as well as building materials, tools and furnishings.

After all participants agree to this virtual construction plan, the construction project is implemented according to this construction plan. Individual construction workers take on tasks within their construction group during the appropriate construction phase. The computer programme for virtual construction planning can automatically select suitable construction workers, tools, building materials and furnishings from the material database and output a duty roster for construction workers and an order list for providers. As soon as construction workers confirm their availability and the providers confirm their availability, the virtual construction project can be implemented in reality. During the construction process, the site managers continuously enter all progress into the profile. These entries can be automated by having construction workers wear Virtual Reality glasses[64] that tell them their work steps, materials and tools and document their progress. Construction workers can view their work steps in the virtual construction project and fast-forward and rewind. An algorithm continuously compares the virtual with the real construction progress.

5.7.1 Assemblies

In the Infrastructure Directory, all the tasks to be done are displayed and the voluntary citizens can join the corresponding assembly groups. For example, roofing would be a construction

---

64 Ministry of Digital Affairs - 13.6.9.1 Virtual reality glasses

group in which volunteers sign up. Volunteers can choose the locations they would like to work on and the period of time they are available. The list of activities is automatically generated and, for state buildings, reviewed by the responsible Building Office and approved or rejected by the electorate during the budget vote[65] . The activities are taken over by the appropriate building groups.

Assemblies consist of at least one construction worker who has the appropriate educational qualifications for the construction section. The programme automatically calculates how many persons need to be trained and how many persons can be unskilled labourers. All matching persons are notified whether they would like to participate in the construction group and which tasks they would prefer to do in which construction section. This data is used to create a digital duty roster[66] . All persons who participate in the construction are registered via their profiles at the Persons, Education and Labour Directories in order to be able to match their place of residence and their interests, educational qualifications and work experience to the appropriate activities at the nearest construction project. Workers from the Planned Economy and Social Market Economy are granted special leave, subject to the approval of their company, to work for the Ministry of Infrastructure.

5.7.2 Material database

The material database consists of all building materials, construction machinery, tools and furnishings that can be purchased or lent out through the Building Yards, the Construction Team, the Procurement Office, Lending Stores and the Central Stores[67] of the Social Villages. The material database lists all available devices owned by the state. This ranges from construction vehicles of the army to vans of the local Building Yards. With this database, the ministries have the opportunity to keep the available resources in operation as constantly as possible, thus contributing to value creation.

---

65 Ministry of Finance - 9.5 Budget vote
66 Ministry of Planned Economy - 7.6.1 Digital duty roster
67 Ministry of Planned Economy - 9.4.7 Central Warehouse

5.7.2.1 Lending warehouse

The Lending Warehouse is a database in the Infrastructure Directory into which goods in storage have been digitised so that they can be lent out. Humans store many things that they rarely need and that can be used multiple times. Through this database, users can use their People's Computers to photograph the items on loan, record the owner's and borrower's identity cards, and regulate damage through liability insurance. Used goods can also be sold or exchanged via the lending warehouse. The transactions are also to be receipted with the People's Computer.

5.8 Construction Team

The Construction Team is responsible for all state construction work that cannot be carried out by a Building Yard alone. It is the central point of contact for orders from ministries and Building Yards. State construction projects are carried out as quickly and thoroughly as possible. To this end, the Construction Team is administered centrally in order to distribute the workforce in such a way that many humans and machines work together on a construction site in a coordinated manner for as short a time as possible, so that no construction project drags on for long. The Construction Team is decentralised into many small units and organised militarily. On the construction site there are four ranks: unskilled worker, construction worker, construction manager and architect, which are sorted in ascending order according to educational qualifications in construction and professional experience. As soon as a large construction project is due, the small squads are pulled together from all over the country with their construction machines. If the construction projects are so demanding that services have to be purchased for a construction project, domestic companies of the Planned Economy, Social Market Economy, Free Market Economy and lastly foreigner companies are engaged first.
The Construction Team is responsible for all state-owned construction machinery, from spades to excavators. In addition

to its own construction machinery, this also includes all things that are not currently used for disaster management, national defence or by the Building Yards.

5.8.1 Workers

The Construction Team consists of permanent professionals, teachers, learners, volunteers, People's Service providers and prisoners. The teachers and learners come from state educational institutions with the relevant school subject or subject area of the colleges. If necessary, specialists from Planned Economy, Social Market Economy, Free Market Economy and foreigners, in that order, are hired on a temporary basis for the necessary construction phase.

Depending on their experience, the workers are assigned to different tasks and responsibilities. If there are persons with previous experience among the volunteers and prisoners, they are given more demanding tasks or authority to instruct unskilled auxiliary workers. All citizens may report for voluntary service to work on state construction works.

The construction workers live year-round with their family in a mobile city that is set up near the large construction project. For smaller construction projects, they live in individual living containers from the mobile city range. Building Yard staff are required to keep a guest room available to accommodate up to 2 construction workers during their construction project away from their homeland. Nearby Social Villages are briefly utilising their capacity to accommodate construction workers. Whenever possible, construction workers are deployed close to home in their municipality's Building Yard. Which construction worker is deployed where is decided in the digital duty roster.

The number of construction workers in a Construction Team depends on the number needed for the construction project. The administration of the construction team is able to administer up to several hundred thousand persons for a construction project in a hurry.

5.8.1.1 Ministry of Education

The skilled craftspersons, engineers and architects are teachers at the state educational institutions. Teachers whose qualifications and specialisation are suitable for this work alternately in the classroom and on construction sites. Students of a craft, engineering or architecture learn theoretical subject knowledge at the college in the winter semester and practise practical implementation together with their teacher on the state construction site in the summer semester. Students of crafts, technology and house construction work on a state construction site for up to 2 weeks per learning year.[68] For large construction projects, all state educational institutions nationwide may send their workers to this one site and live in a mobile city.

5.8.1.2 Ministry of Justice

Auxiliary workers are prisoners without the potential to endanger others, who are also allowed to take on higher tasks depending on their previous experience. They wear collars at all times[69], by which they can be recognised and by which they can be located or numbed if they leave the premises without permission. Prisoners work mainly on construction sites, at night or at weekends, where minors are not present. Prisoners are not allowed to talk to anyone unless it is necessary for official business and they have been given permission to do so by a supervisor. A microphone is installed in the collar that is tuned to the wearer's vote and a measuring device that measures the speech activity in the larynx.

5.8.1.3 Free Market Economy

Technologies and skilled labour may be used from the Free Market Economy for a short time, but in the medium term they are to be researched or trained inland. These external

---

68 Ministry of Education - 9.15.1.7 School subject in the eleventh learning year: house building, 9.15.2.12 crafts, 9.15.2.11 technology
69 Ministry of Justice - 7.5.8 Collars

workers are exceptionally included in the Construction Team and are the most expensive to purchase.

5.8.1.4 Social Market Economy

Specialists from the Social Market Economy are skilled craftspersons or master's craftspersons, graduated engineers and architects who are involved in the construction of infrastructure measures with their companies. These external workers are not part of the Construction Team all the time, but only when their reinforcement or expertise is needed for a construction project.

5.8.1.5 Planned Economy

Auxiliary workers are volunteers with no previous experience from the Social Villages. If someone has previous experience, the caretakers may decide whether they can currently spare the skilled worker or not. Skilled workers are only indispensable if something has to be built or renovated in the Social Village and if the available staff in the Social Village would not be sufficient for this.

5.8.1.6 Barter Economy

From the barter economy come workers who either want to earn money for insurance or who need the short-term use of heavy construction machinery with them for their community, whether in the field or in the capital city to renovate the town hall, school, hospital or supermarket. In this way, they barter by sending construction workers and in return the Construction Team stops at their place to implement their construction projects.

5.9 Mobile city

The mobile city is produced by the industrial community container construction. Mobile cities are used to house the Construction Team, to build a mobile Social Village[70] or to be used as a mobile prison[71] . Other versions are produced on orders from citizens or for export.

A city of containers can be built with the least effort wherever there is a road, railway line, river or sea nearby. Otherwise, a railway line and road are built on top of each other into the area, with the road on the surface and the railway line on stilts or in a tunnel above or below. The containers can be transported via these arteries.

At the installation site, a cross is formed from four containers placed horizontally and at right angles to each other, in the middle of which another container is installed vertically and used as a staircase or lift shaft. This unit is a container cross that serves as a basic module.

5.9.1 Living

When the container crosses are stacked, houses and housing complexes can be built in a grid structure. For this purpose, the vertically placed containers are stacked plumb and serve as stairwells, lift shafts or pillars. The ceilings of the containers can be partially glazed and have a flat or a peaked roof. The sides may consist entirely of panes or windows, which may be mirrored.

One module can already be a flat consisting of a kitchen and dining room, bathroom, living room and bedroom. Two modules can be a single-family house with two storeys, which can be extended if necessary. The wall concept is variable. Via the vertical container, one can access the roofs of the containers and the next floor. On the upper floor there are

---

70 Ministry of Planned Economy - 19 Mobile Social Villages
71 Ministry of Justice - 7.5.5 Prison building

isosceles triangles between the 90° angles as a balcony or as an extension a roof garden. At the outer corners, the containers are connected via the transport holes in all four corners with supporting pillars made of bamboo, on which ivy or grapes climb up. If privacy is desired, plant nets are stretched between two housing modules.

Three or more modules stacked on top of each other form high-rise apartment buildings in which each flat is a storey. The advantage is that the parties do not easily disturb each other with noise. Disadvantage is three, four or five outer walls, depending on whether the containers are placed directly on top of each other or not. Heat or cold loss can be reduced by a tent roof over the grid-like building complex or insulation materials in the walls. The concept can be expanded many times over.

5.9.2 Business

The residential part of the building is on one side of the transport routes, the part with the economic buildings on the other side. The economic wing is divided into commercial, office and production buildings in this arrangement. Next to the road are shops on the ground floor, above them are offices and behind them are production modules; their inner courtyards have glazed ceilings built in to accommodate a factory building. Side elements can also be removed so that only the vertical containers remain as pillars.

5.9.3 Construction Team variant

In the Construction Team variant of the mobile city, 6 short-term construction workers share one bedroom. The remaining housing units are for permanent construction workers who take their families with them to the construction site. The mobile city travels behind the state's major construction projects and moves on as soon as they are implemented. Children go to school in the mobile city and there is a town hall, shopping markets, a hospital and production buildings for building materials that cannot be made in modular construction. In the event of a disaster, the Construction Team is moved to the site of the disaster to rebuild the area. For this purpose, additional beds are placed in all bedrooms to make room for the inhabitants of the destroyed disaster area.

5.10 Infrastructurators[72]

Infrastructurators are giant infrastructure-building machines, each capable of building a road, a tunnel or a house. There are therefore three different types of infrastructurators. These construction machines are either dismountable and thus transportable on roads, rails or ships. Or they are so stable that they can be flown to their operation site by transport aircraft or are able to fly independently and take off and land vertically.
An industrial community is formed for the production of infrastructurators and a People's Innovation Company is established for each infrastructurator.[73] The industrial community includes domestic companies, all domestic defence companies that only switch to defence industry in case of war, and many other domestic manufacturers of construction machinery. The production sites are built in a structurally weak region where, at best, structures of past industrial plants can still be used.

5.11 Container construction

The industrial community for container construction produces all containers and container modules that are used for disaster management, mobile Social Villages and cities, prisons, the Construction Team or for other consumers. Production for other consumers is only permitted if this does not cause a shortage for the state supply. The industrial community for container construction mainly uses used containers from deep-sea shipping and converts or upgrades them.

72§226.6 Housing and home ownership promotion: KV Art.40
73Ministry of Innovation - 13 People's Innovation Company without more detailed description

5.12 Building materials[74]

The infrastructurators are designed to install standardised prefabricated building components, from bridge piers to bathrooms. These building materials are brought to the infrastructurators' operation sites by a network of providers. All domestic building material producers combine their knowledge, skills and capacities in the industrial community for building materials to design and produce these new standard products on the assembly line. The aim is to produce high quality, durable and reusable goods. The price reduction comes from the shorter construction time and automated assembly line production of things like bathrooms, tunnel walls or bridge girders.

5.13 Housebuilding programme[75]

The Ministry of Infrastructure runs the housebuilding programme to raise and maintain the home ownership rate to 90%. This promotes housing construction and the possibility for nationals to buy their own flats and houses in which they live. A population that has home ownership does not have to spend its Unconditional Basic Income[76] on rent. In principle, finding housing is the responsibility of the citizens themselves, and the construction of private buildings is at their personal discretion. However, the Ministry of Infrastructure supports the citizens if they want to do so. Then the people looking for housing jointly choose the locations of their new homes, contribute their own ideas, which they pay for themselves, and work on the construction site as much as they can. Single- and multi-apartment houses with gardens and balconies are built in new housing estates. The buildings with flats are designed for up to 20 persons. Construction workers on the building sites of the housebuilding programme are the new owners and the Construction Team together with their voluntary helpers. The housing units are eventually transferred to their owners through lease-purchase or after direct payments.

74§226.6 Housing and home ownership promotion: KV Art.40
75§45,1e Welfare state: BV Art.41, §198,1 Home ownership: KV Art.40, §226,1,2,4,5 Housing and home ownership promotion: BV Art. 108
76Ministry of Finance - 6 Unconditional Basic Income

The aim of the housebuilding programme is to obtain sufficient building land through the Building Office, to apply standardised procedures in construction and contracting, and to reduce building costs. Thanks to low construction costs, all income groups can afford home ownership as well as reduce the share of housing costs in a person's total expenditure. The buildings are all built according to the latest standards of resource efficiency, which further reduces ongoing housing costs.

The housebuilding programme also serves the purpose of enabling citizens to build a house and maintain their own home in a professional manner. For this purpose, children are trained in housebuilding and craftspersons at school and spend several weeks on the construction sites of the housebuilding programme for practical training.[77] All voluntary citizens have the opportunity to participate in the Construction Team on a voluntary basis and gain first-hand experience.

5.13.1 Granting

First, the housebuilding programme supports families with children with housing, then the low-income, the disabled and finally pensioners. Then it is the turn of all the remaining persons who have submitted motions for home ownership to the Building Office in the town hall. The date on which the application was submitted decides whose turn it is. At least 20% of the buildings are sold with priority to residents of the city who have lived there for at least 10 years and are also looking for a new place to live. This is to facilitate the integration of the newcomers.

5.13.2 Site selection

The locations where the housebuilding programme is deployed are selected according to demographic, democratic and personal criteria. As soon as all locations are occupied and all building types have been determined, the Construction Team moves in and builds roads and houses in series with the help of the infrastructrructurators.

77 Ministry of Education - 9.15.1.7 School subject in the eleventh learning year: house building, 9.15.2.12 crafts, 9.15.2.11 technology

5.13.2.1 Demography in the building area

The demographic viewpoints are derived from statistical values on high-rise residents, the housing market and involuntary tenants. High-rise residents are measured as the number of residents who live there with children. For the housing market, housing shortage is when demand exceeds supply by at least 10%. In the case of involuntary tenants, their details are recorded in the Real Estate Directory or their motions for home ownership are recorded at the Building Office. Users of the Real Estate Directory can indicate in their status whether they are only renting because they cannot afford to own a home on their income, but would like to continue living in the city. The same selection option is available in the application for a home of one's own.

5.13.2.2 Readiness of the local municipalities

This demographic data is used to evaluate locations in whose vicinity new residential areas or satellite towns should be built. All municipalities in the vicinity that are no further than 20 kilometres away and have good transport connections are asked to vote. This brings the democratic aspect into play. Each municipality must vote with its citizens whether, where and how fast it should grow or shrink. A maximum number of new inhabitants can be set in the voting. Municipalities next to which a satellite town is to be built also vote on this and can set an upper limit for the size of the satellite town in terms of area.

5.13.2.3 Selection by the future owners

Based on this democratic data, willing municipalities are marked in the Real Estate Directory. Participants in the housebuilding programme now choose the location for their future home in one of the marked municipalities. Up to 3 municipalities can be selected. In addition, the number of persons with whom one wishes to live in a single or multi-party house must be specified. Voluntary entries include whether one would prefer to have unmated persons, families or pensioners in one's neighbourhood. Via a data retrieval, all interests, educational qualifications and professional experience can be matched in order to find the most suitable neighbourhood.

For locations with sufficient demand for a housing estate or satellite town, the municipal Building Office expels building areas in the Real Estate Directory so that future owners can choose their place of residence. The building plots already indicate which type of building is to be erected here. The users who have made the same election are notified and can come to an agreement with each other. If they cannot agree, the decision is made by lot. The most popular locations are allocated first by the Ministry of Infrastructure. There is no entitlement to a particular location, but there is the option to forego it or to agree on another type of building at this location that offers space for more inhabitants in one place.

5.13.3 Financing

The housebuilding programme is financed by real estate bonds on the People's Stock Exchange as well as by down payments, direct sales and hire-purchase procedures of the future owners. Through the direct sales, real estate bonds are redeemed and through the current revenues from the hire-purchase procedure, interest is paid on the real estate bonds. As soon as the application for a home is made, it is asked how much the down payment should be. In addition, a building savings contract can be taken out to provide a regular fixed amount as a down payment. It can take a while until the desired location is in the queue. The time remaining until completion is shown to the down-payers via the Real Estate Directory. During this time, the money is invested in real estate bonds of the housebuilding programme and their interest is used again as another down payment.

Those who participate in the construction of their own home receive a discount per person who works free of charge and can prove sufficient construction progress in the Infrastructure Directory. Those who continue to participate voluntarily in the Construction Team can earn further discounts.

5.13.4 Sale

The Ministry of Infrastructure sells properties from the housebuilding programme only to private individuals and owner-occupiers, not to landlords and under no circumstances to joint-stock companies. As soon as a house is completed, the purchase price is paid all at once or paid off in the hire-purchase procedure. The annual interest rate included in the hire-purchase procedure is adjusted to the inflation rate of the Social Market Economy's currency. Regardless of which method is used to pay for the homes, the purchase price is equal to the construction costs plus 10% profit.

5.13.5 Resale[78]

The properties from the housebuilding programme may be exchanged at any time and rented out after at least 10 years in owner occupation, but not sold for the time being. They must be bequeathed to the descendants. All descendants are entered in the land register. Their entry is deleted as soon as they own their own home. The house must be rented out if at least one child does not own a home. The rental income that exceeds the maintenance costs flows equally to all children into a building savings contract at People's Bank[79] . The house may only be sold as soon as the last descendant has to leave the land register because he or she owns a home. If the parents have already been able to finance homes for all their children through their income, the parents may sell the house. The sale may only be made to nationals until the home ownership rate has reached 90%.

It is also possible that the home from the housebuilding programme is exchanged for the home of another participant in the housebuilding programme. As all these houses are standardised and built by the Construction Team of the

---

78§227,1,2 Rental business: BV Art. 109
79Ministry of Finance - 11.11.2.1 Building savings contract

Ministry of Infrastructure, equal material and quality is exchanged.

5.13.6 End of the programme

Once 90% of the nationals own a house, whether they live in it or rent it out, the housebuilding programme is stopped. After that, the Construction Team will demolish old housing estates and build new homes, which will be offered to all citizens on a rent-to-own basis. The houses will then be energy self-sufficient and have enough land to grow crops and raise livestock for their own use. Housing in large cities is then reserved for companies and private individuals. Exceptions are the Social Villages, which can also be located near big cities.

6 Networks[80]

The Ministry of Infrastructure provides the networks for pedestrians, vehicles, trains, ships, aircraft, electricity, data, drinking water, sewage, liquid or gaseous fuels and waste. The networks are merged as far as possible. The networks for electricity, data, drinking water, sewage and waste are always laid together and under or next to roads and railways. They support both private households and companies. Households and companies that cannot support themselves with sufficient energy are also provided with pipelines for liquid or gaseous fuels.

The existing network is being optimised so that new mobility concepts can transport more, more cheaply and in a more environmentally friendly way and reduce the fragmentation of the landscape. Optimisation will take place as soon as innovations such as the infrastructurator earthworm and turtle and the flying car are ready for series production.[81] Within the next 50 years, motorised passenger transport is to be shifted to the air. On the one hand, through rails on pillars and, on the other, through a digitalised airspace. Motorised freight

---

80 §200,4,5 Traffic: BV Art. 81a, §206,1,2 Transport of energy, data, water and waste water: BV Art. 91
81 Ministry of Innovation - 13 People's Innovation Company without more detailed description

transport will be moved below the earth's surface by means of rails in tunnel tubes with vacuum.

6.1 Marketing

The Ministry of Infrastructure plans, builds and operates the networks in voting with the people. The National Building Office takes care of the planning in cooperation with the affected municipal building offices. Construction and maintenance are carried out by the Construction Team and the municipal Building Yards. Operation is taken care of by the State Utilities and Municipal Utilities Company, which may only use renewable energies for this purpose. The grid agency is responsible for leasing. The grid agency enables private individuals to use the networks via the Municipal Utilities Company and companies via direct marketing. It sets quotas and fees. The quotas prevent grid congestion and the fees enable profitable operation that can finance the expansion. The networks market is regulated in order to conserve resources and the environment. Networks are a classic natural monopoly. Building the line costs a lot once, but supporting one more person on an ongoing basis costs little. Moreover, it is uneconomical for every private supplier of data, energy, sewage or transport to lay its own line. It is better for one supplier to lay all the lines in bundles and ensure sufficient capacity to support the country adequately. Since the lines support the population with essential services, this monopolistic supplier must be state-owned so that the people can control it democratically, for example to avoid excessive prices or negative externalities.

The price for the use of the lines is calculated from the total average costs, the external costs and 10% profit. The total average costs consist of the construction costs, which are paid off pro rata, and the current operating expenses. Once the construction costs are paid off, the price decreases. The external costs consist of the pollution left behind by the users of the networks and its disposal. Innovations and laws reduce the pollution, which then results in a price reduction. The price is set for a certain quantity and level of wear and tear. The fees to be paid depend on the quantity used and the resulting damage to the networks and the environment, which must be repaired by the network operator.

6.2 Underground networks

The Ministry of Infrastructure lays underground networks. They run about 20 metres underground as a tunnel tube with a diameter of 15 metres, so that it is covered by 5 metres of earth. Inside this large tunnel tube run 4 tubes for transporting solid materials on rails. The tubes each have a diameter of 4 metres and are arranged at right angles to each other. This means that there are always two tubes above and next to each other. The traffic in the tubes above each other goes in the opposite direction. The traffic on either side can turn right or left. In these tubes, there is a vacuum in which trains run automatically. The bottom plate of each wagon contains the drive technology of the Maglev. Wagons can be coupled to the train from the front or rear while it is moving.

Next to these 4 pipes run 5 pipelines with a smaller diameter than the pipes and a separation in the middle. Drinking water and data flow in the first pipe, liquid and gaseous fuels in the second, waste water and an empty pipe for upcoming innovations in the third, salt water from the sea flows in two opposite directions in the fourth and electricity and cooling liquid flow in the fifth. The cooling liquid surrounds the power cable and conducts its waste heat into a heat exchanger to generate electricity.

Rescue and maintenance shafts run between the tubes and pipes, as well as a grid structure that can flexibly compensate for a displacement of the large tunnel tube due to tectonic shifts without causing damage to the individual tubes or pipes. In order to support a settlement, pipes for electricity, sewage, possibly fuels, data and drinking water are branched off and laid under the streets of the settlement up to the buildings. The solid materials are stored in containers that are transported on their wagon via feeder pipes from the transport pipes of the underground network to the underground freight stations of a settlement. The components for the underground network are prefabricated and installed in standardised procedures. The infrastructurator turtle builds the underground networks over land and the infrastructurator earthworm builds them through mountains or under settlements.

The route is being started in several places at the same time, so that about 100 infrastructurators are in use. As long as sufficient electricity is not yet available through constructed facilities, the "blue whale" infrastructurators support the work with electricity. Above the tunnel tube, solar power plants are being built over land to support the route with energy. In polar regions, wind turbines can be built. Pumped-storage power stations will be built on mountains. The tunnel tube runs in a grid from north to south and east to west. The people decide how close-meshed the underground network should be, i.e. whether there should be only one or several tunnel tubes per cardinal direction.

6.2.1 Electricity network

Power pylons over populated areas are being removed and underground cabling is being laid. In the short term, electricity pylons are erected in order to be able to transport sufficient electricity from wind and hydroelectric power plants at sea to the interior of the country. In the medium term, the north-south Maglev route and the tunnel tube including the power line will be ready and many electricity pylons can be dismantled. Electricity pylons that can be retrofitted with a wind turbine to generate electricity and that a majority of citizens in a municipality want to keep in operation can be left standing. In the long term, electricity will only flow in underground lines to make use of its waste heat.

The electricity network is rented to all electricity producers who feed electricity into the network. The rent is borne by the sales price. Private individuals who want to feed in excess electricity from their own generation must register a company for this and purchase an electricity meter. Electricity fed in from renewable energy sources receives a lower rent. Electricity from fossil energy sources is charged a higher rent, with the increase going towards the expansion of renewable energies.

6.2.2 Fuel network

The fuel pipelines are used to transport fuels in liquid or gaseous form, which are used the most. In the short term, this is crude oil and natural gas, and as soon as possible bioethanol and hydrogen. The pipelines are leased to the producers of the fuels.

6.2.3 Data network

The broadband network consists of 5 lines, namely radio, television, telephone, internet and intranet. The intranet line is specially secured and may only be opened or switched in the presence of an employee of the Ministry of Intranet. The lines for radio, television, telephone and internet are leased to telecommunication companies. If new technology becomes available, more cables can be laid in the conduit. No cables are used to reach the end user. This is done via transmission masts. The Network Agency is responsible for the operation and administration of these radio frequency-based networks. The network agency leases and sells radio frequency ranges and monitors that only these frequencies are used.

The data network is leased to suppliers for internet, television and telephone. The suppliers can choose which frequency ranges they want to rent, how high the transmission speed and the total amount of transported data should be. The higher both are, the higher the rental price. The rental fees are used to finance the maintenance of the data network for the intranet.

6.2.4 Sewage network

The sewage network runs through all cities and is known as the sewer system, which flows into a wastewater treatment plant. In addition, the sewage network runs through the entire country and is capable of pumping or draining water to every region of the country. This wastewater is treated sewage or river water, optionally also seawater. The sewage network is connected to all rivers and coasts inland. It provides the basis for all pumped-storage power stations. Local suppliers are the respective Municipal Utilities Companies.

The wastewater network is paid for by a fee per litre of wastewater according to the water meter at the inflow to the treatment plant. The fee includes the cost of treating wastewater to drinking water in treatment plants and the research costs to be able to dispose of all residues. The manufacturers of the products that end up in wastewater have to pay the fees. For example, a pharmaceutical company that sells the contraceptive pill has to pay for research into filter mechanisms for female hormones in drinking water and the nationwide installation of the filters in all sewage treatment plants through the wastewater fee.

6.2.5 Drinking water network

The drinking water network is fed from the sources of the Municipal Utilities Company and is subject to strict control by the health auditors of the Company Auditing Agency.[82] This ensures that drinking water quality is similarly good throughout the country and that drinking water can be provided over long distances during periods of drought. Local suppliers are the respective Municipal Utilities Companies.

The drinking water network is paid for via a fee per litre of drinking water according to the water meter located at the inflow to a household. As long as drinking water can be extracted from the ground without treatment, only the pipe costs are incurred by the end users. Companies that want to use fresh water that is not intended to be drunk must use industrial water or pay 3 times the price for a litre of drinking water.

6.2.6 Salt water network

The saline water network is primarily used to support underground pumped-storage power stations in the mountains. However, it can also be used to fill purifiable reservoirs and let the water evaporate to produce fresh water.

82Ministry of Labour - 20.7.2 Health auditor

When water is scarce, it can be used as industrial water.

6.2.7 Waste disposal network

The local sewerage system is extended for waste disposal. Waste containers are placed at sewer manholes to which the residents bring their waste. These containers have several openings for different kinds of waste. In the container, the waste is pressed into shape and packed into transport containers. Each transport container has a different colour, depending on the type of waste. The transport containers are lowered into the sewage system. On the ceiling of the sewage system there is a system of pipes for pneumatic tube mail through which the containers are transported to the treatment plant. At the treatment plant, the transport containers are emptied, cleaned and automatically sent back to the waste containers. The waste is picked up at the treatment plant by the waste collector and transported to the appropriate waste recycling plant. Should waste be transported through the underground network, it is loaded into containers.

Funding for the operation of the waste disposal network and research into the prevention of waste generation is financed through fees. The fees for the waste disposal network vary depending on the recyclability and toxicity of the waste. They are paid by the producers of the disposed products or, in the case of imports, by the importers.

7 Digital infrastructure

The digital infrastructure is planned by the Ministry of Infrastructure in voting with the Ministry of Digital Affairs. In the planning process, the strategic aspects of digitalisation are determined in voting with the ministries of labour and economy. This identifies the current and future demand for data that is transmitted by cable or via radio signals. For radio signals, the network agency operates a frequency policy that shields the intranet from the rest of the radio traffic via special radio masts and frequencies. All remaining frequencies are auctioned off in bidding procedures to companies for mobile radio every 5 years. The revenues are used to make technological innovations to the radio masts.

The digital network infrastructure is built by the Construction Team in cooperation with the People's Innovation Company Intranet.[83] Where the construction activities take place is determined by the Ministry of Infrastructure in voting with the Ministry of Digital Affairs and the people in the digital agenda, which must be formulated at a committee and confirmed in a voting. The digital network infrastructure is maintained by the State Utilities and audited by the technical auditors.

7.1 Frequency monitoring

In general, all devices that are broadcasters or receivers for protected frequency ranges must be registered with the network agency by the seller or importer. Monitoring works via an encrypted transmit command. Only devices that are registered to use this frequency recognise the command and each sends its own recognition code to the network agency. This is done via the broadcasters, which are located all over the country and transmit signals for radio, television, mobile internet, telephone and intranet.

Broadcasters and receivers operating in the frequency range of the intranet always also transmit and receive malware or a virus, which they can only ward off if they are authorised to use the intranet. The Ministry of Digital Affairs ensures implementation.[84]

To further monitor data security, frequency control vehicles and aircraft are used to check frequency use on a random basis. The frequencies on which the People's Computers transmit their signals are monitored particularly closely and regularly. Mobile communication devices from Planned Economy and Social Market Economy are monitored on a random basis.

The background is that frequencies can be manipulated through third-party use. For example, data could be fed into the intranet that a user is not even authorised to feed in. For

---

83 Ministry of Digital Affairs - 13 People's Innovation Company Intranet
84 Ministry of Digital Affairs - 8.2 Cyber Defence

example, this could be an entry that only the Ministry of Digital Affairs is allowed to export. Or the intranet could be bugged in order to collect data, but access to which would have to be authorised by the user.

Exceptions are permitted in a cleared frequency range which the network agency defines and does not monitor. Devices that are only receivers of free-to-air signals, such as radios or televisions, have no function for encryption.

8 Traffic[85]

One of the tasks of the Ministry of Infrastructure is transport, which opens up all areas of the country. The necessary transport infrastructure should always be sufficient to ensure that all persons and goods can move forward without unnecessary delays and congestion. It should also be up-to-date to ensure the highest possible speed and environmental compatibility. To this end, the Ministry of Infrastructure facilitates environmentally friendly individual transport as well as state and private local and long-distance transport.

In cooperation with the driving and aircraft industries, land, sea, air and space vehicles are being imported that have new environmentally neutral propulsion technologies and refuelling options. The Ministry of Infrastructure is enabling automated and connected driving and flying through novel transport routes and appropriate regularisations. The Traffic Office is responsible for digital land, sea, air and space management. The Company Auditing Agency's technical auditors check the safety of these new technologies and the innovation auditors support the introduction of new technologies.

8.1 Transport law

In the area of transport, the Ministry of Infrastructure coordinates its laws with the Ministries of Labour and Safety. Transport law regulates which persons may be transported with and without luggage and which goods may be transported

---

85§200,1,2,3 Traffic: BV Art. 81a, §201,5 Road traffic: KV Art.34, §202 Transit traffic: BV Art. 84

within the country and across borders.[86] In the transport law, the Ministry of Infrastructure regulates that goods may only be transported by rail as long as they are transported on land and until transport by rail is no longer possible to reach the customer. Commercial transit traffic through the country must take place either by rail or by air.

8.2 Traffic law

Traffic law is determined in voting with the ministries of health, safety, innovation and labour. The Ministry of Infrastructure regulates the individual means of transport, their route planning and digitalisation in traffic law. The requirements for environmental protection are regulated in voting with the Institute of Environmental Medicine[87] . Requirements for road safety are regulated in cooperation with the institutes for technology and occupational health as well as the technical auditors[88] . Traffic law is divided into traffic on roads, railways, waterways and oceans as well as in airspace and outer space.

The People's Protection Service[89] and the Company Auditing Agency technical auditors are responsible for checking compliance with traffic law.

8.3 Traffic Office

For the execution of traffic law, the Ministry of Infrastructure operates the Traffic Office. It has responsibilities in all areas of transport and operates offices that specialise in their respective areas of transport, namely the Road, Rail, Water and Air Transport Offices. As the central agency, the Traffic Office provides statistics and forecasts for the transport of persons and goods. It records how many persons and goods use what means of transport and when. Based on the data, it pursues

---

86Ministry of Labour - 20.7.6.4 Compliance with trade law, 10.3Foreign trade regulations, Ministry of Security - 8.2 Border management
87Ministry of Health - 4.5.4 Institute of Environmental Health
88Ministry of Innovation - 7.1 Institute of Technology, Ministry of Health - 4.5.5 Institute of Occupational Health, Ministry of Labour - 20.7.4 Technical auditor
89Ministry of Security - 6 People's Protection Service

an investment policy in state transport systems and means of transport for persons and goods. It issues regularisations for the transport of dangerous goods and for the transport of persons. Persons can contact the Traffic Office with concerns about consumer protection in the area of transport.

The Traffic Office regulates which areas of knowledge and physical abilities are necessary for driving the respective means of transport and how these requirements differ from persons who professionally transport persons or goods or professionally undertake driving licence training. Theory classes may be held in state educational institutions.

The Traffic Office, in cooperation with the Company Auditing Agency's technical auditors, is responsible for investigating accidents caused by means of transport on the roads, railways, waterways and in the air. This involves investigating structural deficiencies or faulty planning that promoted or triggered an accident.

8.4 Environmental protection in transport

Environmental protection in transport includes keeping the air clean, avoiding noise and keeping out waste and exhaust fumes. Mobility in cities is therefore provided by muscle power, electricity or hydrogen. For electric mobility, the Ministry of Infrastructure prescribes a standard for exchangeable batteries for means of transport, which are changed at filling stations so that the cars can continue driving directly. The charging process can take place at the filling stations or at any charging station at the car park. Filling stations are obliged to offer hydrogen in addition to fully charged exchangeable batteries. From 2040, petrol stations will no longer be allowed to sell petroleum-based fuels. From 2045 onwards, petroleum-based means of transport will no longer be allowed to travel inland, to domestic ports or to fly over domestic ports.

The internal combustion engine and combustion jet engines are being replaced by alternative fuels or electric drives, which is being used as a noise abatement strategy and immediate clean air programme.

8.5 Digital mobility

Digital mobility enables means of transport to move autonomously, to take care of each other and to reach their destination as quickly and safely as possible. To achieve this, means of transport communicate with each other and with the Global Navigation Satellite System via radio. The Global Navigation Satellite System consists of many satellites in orbit that are used for earth observation. They collect all the data on the weather and the current location of all means of transport. The Ministry of Infrastructure receives the meteorological data from the Meteorological Service and the data of the means of transport via the Global Navigation Satellite System. The exchange of data between the means of transport takes place via open data that the vehicles share with each other.

Through the Global Navigation Satellite System, means of transport are able to recognise each other, avoid each other and communicate with digital traffic guidance systems. The traffic guidance systems ensure orderly traffic without accidents and the fastest possible progress for all road users. Artificial intelligence in autonomous means of transport may not be used on roads in cities. In voting with the people, the Minister of Infrastructure shall enact laws that include requirements for autonomous means of transport with artificial intelligence. Artificial intelligence transport shall be reserved for long-distance transport, water and air.

8.6 Transport Directory

Citizens can use the Transport Directory to form carpools and view all timetables and schedules. Every regular transport line has a profile here and every one-off transport service is given a short-term profile. Entrepreneurs who transport goods or persons form a group with subgroups for each individual company. State transport companies also form a group with subgroups for each individual transport company or transport association. Private or business travellers can form ride and transport pools. Digitally networked vehicles can calculate identical routes and users receive a notification that they could form a carpool because they have the same destinations. Users can search for these carpools themselves or be automatically informed via a data retrieval and share their data with other users, when they move where and whether they do so regularly.

8.7 Road network[90]

Roads connect houses, residential areas, industrial zones, cities, municipalities and countries. In the long run, the road network will always remain the artery of the Ministry of Infrastructure towards the consumer, because all important supply networks run under it.

Pedestrians are given pavements, cyclists are given cycle paths and motor vehicles are given lanes. There are these three types of roads, which can also occur in twos or threes on a road. If they occur on a road, they must be separated from each other by markings or signage. Motor vehicles are not allowed on footpaths, cycle paths and hiking trails. No pedestrians or cyclists are allowed on motorways and highways. Footpaths, hiking trails and cycle paths shall be upgraded. Motorways and highways are to be dismantled and replaced by maglev trains and aeroplanes.

The national road network is planned, operated and administered by the State Utilities, the municipal road network by the Municipal Utilities Company. Road network planning is done in voting with the national or municipal Building Office and in voting with the affected population. Cross-border roads are planned, operated and administered in voting with the neighbouring country. The Ministry of Foreign Affairs is responsible for the negotiations. The Traffic Office forwards the revenues from tolls to the State Utilities or Municipal Utilities Company according to their use.

The road network consists of a pavement of stone, asphalt and concrete, but is increasingly being replaced by solar panels that can illuminate, heat and generate electricity on the road.[91] Wind turbines are installed on roads where vehicles travel at

---

90 §201,1,2,3,4,6,7 Road traffic: BV Art. 82, 83, KV Art.34, §204 Foot- and hiking trails: BV Art. 88
91 https://www.solmove.com/

speeds of at least 50 kilometres per hour outside built-up areas.[92] Street lighting has motion detectors, provided this does not disturb residents. In addition to the road surface, road traffic technology also consists of bridges, tunnels, rest areas, street lights, reflective, illuminated or digital signs, barriers, roundabouts and traffic lights. The safety of the road infrastructure is ensured when the roads are in a condition that does not pose a hazard during normal use. Road traffic telematics is digitalised in such a way that traffic guidance systems prevent congestion and alert the People's Protection Service as soon as road users brake or endanger other road users.

8.7.1 Road traffic

In road traffic, there is traffic on foot, by bicycle or similar means of locomotion powered by muscle power, as well as traffic by motor vehicles powered by a motor. For pedestrians and cyclists, road use is free of charge and only the road traffic behaviour law applies, such as paying attention to traffic signals like traffic lights, signs or zebra crossings, moving on the right side of traffic lanes and overtaking on the left side. For all motorists, the Road Traffic Act[93] applies in full and additionally the Road Traffic Law[94] , which regulates the admission of persons and vehicles to motorised road traffic.

The road network is controlled by the People's Protection Service.[95] The controls are random checks of a valid toll sticker, motor vehicle insurance, inspection sticker of the Company Auditing Agency technical auditors on each motor vehicle, driving licence and identity card of each driver. In addition, compliance with the Road Traffic Act and Road Traffic Regulations is controlled. For example, speed checks are carried out or tickets are handed out for wrong parking or spitting on the road.

92http://devecitech.com/
93https://www.gesetze-im-internet.de/stvo_2013/index.html
94http://www.gesetze-im-internet.de/stvg/
95Ministry of Security - 6 People's Protection Service

8.7.2 Vehicles

Vehicles are approved for road use by the Traffic Office. The dealers or manufacturers must have this approval checked as an individual test by the technical auditors of the Company Auditing Agency. This approval process is repeated in a simpler form every 2 years. It serves the purpose of quality assurance and market surveillance. During this process, the vehicle technology is checked for its durability until the next test and the general vehicle safety is determined. In case of deficiencies, the vehicle may lose its registration until the deficiencies are remedied. The technical auditors also take the driving licence examinations, which must be passed for motor vehicles with a speed of 30 km/h or more. Anyone who is of age of majority may take a driving test.

All newly registered vehicles must not cause environmental damage. They must not emit harmful exhaust gases, no matter in what concentration. The abrasion from their tyres must be naturally degradable. No liquids may be used to operate the vehicle that are not naturally degradable, because even the drop-by-drop exit must be excluded. Manufacturers must prove that their vehicles produce as little noise as possible from 30 km/h.

All first-time registrations of new vehicle types must demonstrate innovative technology when presented to the Traffic Office, technical auditors and innovation auditors. A modified design is not an innovative technology as long as no beneficial effect occurs.

8.7.3 Public transport[96]

The roads are predominantly used by individual traffic. Public transport is provided by bus and taxi companies. Buses run everywhere where no other state transport is offered. If no company can be found that wants to operate the bus lines, but the majority of citizens are in favour of having a bus line, the Ministry of Infrastructure takes over the state operation. Operators are not allowed to use polluting vehicles, their

---

96§201.5 Road traffic: KV Art.34

drivers have to take advanced driving tests and repeat them every 5 years.

8.7.4 Freight

Freight transport by lorry is only permitted up to a total weight of 4 tonnes. These transporters serve the short distance from the nearest railway station to the consignee or from the broadcaster to the railway station. In the transporters, new goods are brought to the consignee and at the same time reusable empty packaging and packaging waste are collected. At the broadcaster, reusable empty and new packaging is delivered and new goods are picked up. Commercial empty runs of transporters are punished with a fine of 100 Dollars.

8.7.5 Highway

The minimum speed on highways is 130 km/h, exceptions are danger spots, construction sites, hard shoulders, entrances and exits. On the right lane is the autopilot lane, where self-driving cars form an energy-saving convoy with a small distance between them, are not allowed to overtake and are allowed to drive at a maximum of 200 km/h.
The age of the highway will end as soon as there are enough flying cars on the roads and flying for the long haul instead of using the highways. The air network will then be responsible for this. Highways will then only be left between Social Villages and otherwise converted into cycle paths, free play areas for the people or demolished and turned into building land or farmland.

8.7.6 Financing of the road network[97]

The road network is financed by a toll for domestic citizens and foreigners. It applies to all vehicles with a total weight of 100 kilograms or more with which the road network is to

---

[97]§200.5 Traffic: BV Art. 81a, §201.8 Road traffic: BV Art. 82, §203 Heavy vehicle fee: BV Art. 85

be used. The use of the roads and paths is free of charge for pedestrians and cyclists and is financed by the toll.

For domestic citizens, the toll is collected annually via the compulsory motor vehicle insurance because it has to record the necessary data for a vehicle. This includes the exact type designation of the vehicle, its age and the annual mileage. From this, it is calculated how much this vehicle wears down the road and pollutes the environment through exhaust gases and waste products. For example, vehicles with rubber tyres produce abrasion that generates microplastics and fine dust. Cleaning the environment from these waste products is costly and needs to be researched. This makes for high tolls on such vehicles due to disposal and research costs. Domestic citizens receive their toll sticker from their motor vehicle insurance company annually by post.

The fees collected are spent in full on road maintenance, state waste disposal operations and corresponding research projects[98]. However, as soon as an innovation prevents abrasion or makes it environmentally compatible, the price surcharge for these vehicles is waived. The Company Auditing Agency's technical auditors classify vehicles accordingly during testing and registration in voting with the health auditors.

For foreigners who cannot provide the above-mentioned proof, there is a toll sticker that can be purchased for any period of time at petrol stations and customs offices. The price is calculated per day.

8.8 Rail network

The rail network is used by trains that are powered by electricity or hydrogen and transport persons and goods. All railways are digitised and networked in the Transport Directory. They form the state transport system, which is operated on land. Private suppliers can operate the rail network or the railways if this can generate additional revenue for the state and public mobility does not suffer. In urbanised areas, the rail network consists of tramways, cableways, underground railways and rapid transit railways for local transport and railways for long-

98 Ministry of Innovation - 5.4.4 Transport transformation

distance transport. Railways consist of railways until they are replaced by tunnel railways for freight transport and Maglev railways for passenger transport.

8.8.1 Local traffic

The railways for local transport in urban areas and their surrounding areas are operated by the Municipal Utilities Companies of the participants municipalities. They transport persons and goods as close as possible to their destination. Persons can get on or off at stations. Cable cars in cities can disturb the privacy of residents who live below or directly next to them. Cable cars whose lines are affected by this must be equipped with window sills that do not allow direct looking down into gardens and windows.

Goods are unloaded from the tracks onto the road at level crossings and vice versa. To do this, each goods train carries a U-shaped trolley that rolls over the container train until it reaches the desired container. There it anchors itself to the wagon, widens until the trolley reaches over the low loader and extends feet to the ground. The container is lifted from the wagon onto a truck or vice versa via two winches and hooks with the trolley.

8.8.2 Long-distance traffic

The mainline railways are operated by the State Utilities. They transport persons and goods as quickly as possible to the desired local transport network. The State Utilities are responsible for switching freight transport from road to rail and for switching from railways to tunnel and Maglev railways. Electrification of existing rail networks will be discontinued and railcars will be converted from diesel propulsion to hydrogen propulsion. No new tracks will be laid, but existing railway lines will be built over with U-shaped piers on which rails for maglev trains will be laid.

A tunnel tube is laid in the ground between the piers of the maglev train. Wherever the Maglev is laid over existing rails, the height of the pillars must be sufficient for trains to still be able to run.

National rail investment will be directed first to the construction of the Maglev system and then to the construction of the tunnel tracks. Once the Maglev networks are sufficient to shift long-distance passenger traffic from rail to Maglev, investment will be made in the tunnel tracks to shift long-distance freight traffic from rail to tunnel tracks. All railway tracks are dismantled, melted down or sold.

8.8.3 Maglev train

The magnetic levitation train, also called Maglev, makes it possible to cross the landscape quickly, quietly and cleanly. It produces no abrasion from tyres or brakes, no rolling resistance and does not cut up the landscape because it runs on pillars. Within 5 to 25 years, Maglev lines run across the country on piers that have an inverted U-shape. This way, persons can travel on the top and containers can be transported suspended between the pillars, also using Maglev technology. There is still 3 metres of space between the bottom edge of the container and the ground, so that persons, animals and vehicles can walk or drive underneath.

The tracks of the maglev train are laid in two lanes above and below the cross struts of the piers. For this reason, the piers are U-shaped because the lower two tracks run between the piers. The lower track hangs on the rails and does not rest on them like the upper track. Solar cells are laid on the top of the rail. Vertical wind turbines[99] are attached to the pillars, which mainly convert the wind from the railway into electricity and come from the dismantled highways.

The entrances and exits are parallel sections to the right or left of the track, which are branched off via a curve towards a station. Before an exit, the last wagon is uncoupled and drives into the exit. After an entry, a new last wagon approaches the train from behind. The last wagons are therefore transit areas, where all passengers stay to get off at the next exit. Luggage is automatically loaded in the underbody and must not exceed

99http://devecitech.com/

certain maximum sizes. Maglev trains always travel at the same speed. They do not stop, only the transit cars.

As soon as the tunnel railway is ready for operation, no more goods are transported on the underside of the maglev train, because from then on that happens in the tunnel. Vehicles of passengers travelling in the maglev train will then be transported there.

In the medium term, a continental Maglev network will be targeted and at least the domestic network will be relocated. Entrances and exits are built on the longitudinal roads, which belong to continental metropolises. Only the main routes are two-lane. They connect the ends of the continent with each other and the metropolises in all directions. This does not mean that the main route will run along there, but the main route will be built as ecological and short as possible. In the long term, this network will be connected to the networks of other continents and span the globe.

Countries that want to participate will be offered an export deal. If countries or companies buy the finished lines and the appropriate infrastructurator, they will receive a corresponding number of trains free of charge on completion.

8.8.4 Tunnel railway

As soon as the first infrastructurator for tunnelling is ready and sufficient tunnel segments have been pre-produced in series by the industrial community for building materials, the construction of the tunnel railway will start. The first section is the north-south connection with a slight gradient from the mountains to the coast. As soon as 10 infrastructurators build the north-south link, more infrastructurators will be used to extend the cities' underground railways further out into the countryside.

The construction of tunnels for urban subways is standardised and consists of a ring that is connected to a cross. In the middle of the cross is the city centre. Similar to a paternoster, the subways travel in circles and allow for transfers at stations. After that, each city is connected to at least one tunnel tube with underground trains.

The tunnels connecting cities criss-cross the country from north to south and from east to west. They are built in consultation with neighbouring countries so that the tubes can later be connected to form a continental and worldwide network of tubes. The tunnels have a gradient from mountains to coasts. The gradient can be used to support the pumped-storage power stations and the natural flow of water can be used by gravity to generate and store electricity. In the four transport tubes, tunnel tracks travel in a near-vacuum using magnetic levitation technology to reduce air resistance. Entrance and exit work in the same way as with the above-ground magnetic levitation train.

In the long term, the number of infrastructurators will determine how dense the Maglev and tunnel railway network is and whether it is capable of connecting the continent and the world.

8.8.5 Financing of the rail network[100]

The rail network with tracks on ballast is sold with the trains. The sale is to companies from the Social Market Economy and Free Market Economy. The revenues from the sale will be used to finance part of the track network for the Maglev.

Annual fees are levied on harmful abrasion from wheels and brakes, which are equivalent to disposal and finance the expansion of the Maglev because it does not require abrasion. The operating licence expires as soon as the maglev train and the tunnel tubes are ready for operation. Transport charges apply for passenger transport in the maglev train and for freight transport through the tunnel tubes.

8.9 Shipping network[101]

The Traffic Office regulates the use of the waterways as traffic routes. The State Utilities operate these waterways and arrange for the necessary hydraulic engineering by the Construction

---

100 §200.5 Traffic: BV Art. 81a, §201.8 Road traffic: BV Art. 82, §203 Heavy vehicle fee: BV Art. 85
101 §192,2,3 Water: BV Art. 76

Team. This includes the construction and maintenance of waterways, harbours, locks, dams, dikes, barrages, flood basins and hydroelectric power plants. In cooperation with the Traffic Office, the State Utilities undertake the infrastructure planning for the waterways and submit it to the Minister of Infrastructure for review.

Interventions in the water cycle must be voted on by the Minister of Infrastructure in a committee with the people. In cooperation with the Municipal Utilities Company and the health auditors of the Company Auditing Agency, water protection is controlled and appropriate measures or limits are prescribed. The necessary measures are carried out by the State Utilities together with the innovation auditors of the Company Auditing Agency.

8.9.1 Waterways

Waterways are all streams, rivers, lakes and coastal areas. The Traffic Office, in voting with the State Utilities and Municipal Utilities Company, determines which waterways may be used for power generation, recreation, sports, inland waterways or maritime navigation. Recreation, especially water sports, is permitted in all waterways. Exceptions apply at bridges, locks and port facilities. The waterway infrastructure consists of harbours, ferries, jetties, locks, barrages, bank stabilisation and beaches. Hydropower is used to generate energy at all suitable locations, at least to support the technology needed for the locks and dams. The Ministry of Infrastructure ensures that freshwater resources such as streams, rivers and lakes are potable through its environmentally friendly management.

Observation systems are used to measure the impact of shipping on the waters and to specify routes for the ships that serve to protect the waters. Water levels and tides are measured regularly in order to guarantee the water supply at all times and to ensure sufficient utilisation of the hydropower plants. The State Utilities in cooperation with the Municipal Utilities Company examine the impacts of shipping and can give the Traffic Office requirements on how to minimise the impacts.

8.9.2 Shipping

The Traffic Office regulates the requirements for the use of waterways for recreation, port management, sports, inland and maritime navigation. Water protection includes keeping the water and air clean and not stressing the banks and bottom of rivers, lakes and seas with currents. Keeping it clean means disposing of all waste and sewage on land and not producing harmful exhaust gases. Waste water can be treated by sewage treatment plants on board in order to be allowed to be disposed of into the waters. All owners of a vessel must pay an annual fee for the use of the waterways, the amount of which depends on the type of vessel and the frequency of use.
State Utilities are responsible for surveying the land underwater and producing maps with the most accurate geological and biological information possible about the underwater environment. They also ensure the operation of wind turbines, tidal and wave power plants at sea.

8.9.3 Rivers

Rivers serve as groundwater suppliers and must therefore be kept as clean as possible. Inland vessels convert their propulsion systems to electricity or hydrogen within 5 years. At the locks or some bridges, empty batteries can be replaced by full ones or hydrogen can be refuelled. In the medium term, inland vessels are to be given air cushions and only enter the water under bridges in order to protect the underwater environment and make faster progress.

8.9.4 Flood basin[102]

Artificial lakes are being created along rivers at suitable locations, especially near large cities. They serve local recreation on the one hand and energy supply and disaster management on the other. They provide fire-fighting water, pumped storage reservoirs, fresh water and flood basins. They have an outlet and an inlet that is connected to a natural river. The weather

---

102§192,1,3 Water: BV Art. 76

service detects severe weather conditions in advance that could bring large amounts of rain and lead to flooding. As soon as a danger is detected, the lakes are drained into the river next to which they were created. When the storm approaches and fills the river, the flooding of the lake is to avoid peak levels that would drive water over the protective dams. During droughts, the outlets can be closed to collect freshwater and pump it through the underground network to affected regions.

8.9.5 Lakes

All lakes in the country are used as gravel pits for construction materials if necessary and possible. All these gravel pits must also be accessible to citizens for local recreation. If necessary, buoy lines are drawn to demarcate swimming areas where dredging is not taking place.

8.9.6 Maritime shipping

Maritime shipping is to convert its propulsion systems to the energy sources wind, electricity and hydrogen. Electricity and propulsion are to be generated by sailing parachutes and kites. With hydrogen, a fuel cell and an electric motor can turn the ship's propellers. The same requirements for water protection apply as for shipping. Sea-going vessels that are not equipped in this way will no longer be allowed to call at domestic ports in 2035.

For the prevention of danger in maritime navigation, the Traffic Office operates the pilotage system and the maritime accident investigation. To ensure safety in maritime navigation, requirements are made for nautical science, technology, radio and seafarers, which also apply to inland navigation as far as possible.

A global navigation network for the oceans ensures safe maritime navigation. It is updated by the Global Navigation Satellite System and supplemented by data from submarine drones and observation stations.

8.10 Flight network

The flight network consists of digital corridors in the airspace in which different types of aircraft move through an autopilot. It extends further into space as time goes on. The Ministry of Foreign Affairs regulates space law in voting with the Ministry of Infrastructure and other states. The Ministry of Infrastructure is responsible for air law within the country's borders. Air law regulates overflights, take-offs and landings of international air traffic, noise, environmental and consumer protection in air traffic, as well as requirements for aviation technology, flight operations, aviation personnel and aviation safety. The Traffic Office is responsible for ensuring compliance with aviation law and can alert the Ministry of Security in the event of violations. For air safety operations, police helicopters or armed army helicopters and aircraft may be used to avert danger. The Traffic Office is also responsible for aviation accident investigations and can notify the police for evidence collection and forensics if criminal acts are suspected. Anyone who is affected by noise or pollution, or who is harmed by the aviation industry as a consumer, is entitled to compensatory damages and, if applicable, damages for pain and suffering. State Utilities are responsible for building and maintaining airports and spaceports, as well as for satellite navigation on Earth and in space. Private and commercial operators are also possible. The State Utilities operate air traffic control with air traffic controllers for pilots and the digital flight corridors for autopilots. This guides aircraft to the desired airports and coordinates them in the airspace.

8.10.1 Airspace

State Utilities hire air traffic controllers to control air traffic in domestic airspace. Different altitude corridors are cleared for different cruising speeds and aircraft types. The airspace is made accessible to individual traffic through the Global Navigation Satellite System with satellites from the Ministry of Innovation and computer programmes for autopilots from the People's Innovation Company Intranet. The Ministry of

Innovation is responsible for the development. For vertical take-off flying cars, suitable parking areas of supermarkets and companies or roofs of multi-storey car parks are selected as take-off and landing sites and released by the air traffic control of the Municipal Utilities Company. The flying cars use these take-off and landing points to reach their cruising altitude, depending on how far they want to fly. The software allows any destination to be approached as the crow flies. There are three options for how cross-traffic is handled by the autopilot. Either one aircraft flies faster and the other flies slower, or an avoidance manoeuvre is calculated that does not cause a collision, even with other aircraft.

Since all aircraft must be connected to the Global Navigation Satellite System, all obstacles, their speed and direction of flight are automatically recorded because you have to set your autopilot before take-off. To do this, one selects one's take-off site and one's landing site. With the help of the aircraft identification, fuel consumption is calculated and refuelling stops are automatically scheduled. Depending on how fast you want to fly, you have to change the battery or refill hydrogen more often. The refuelling stations, take-off and landing areas must meet the safety requirements of the Company Auditing Agency, but are not state-owned. In return for maintaining the parking, take-off and landing area, shops have more customer traffic.

The Municipal Utilities Company provides sufficient free flight zones so that pilots are also able to manage the aircraft themselves and fly for pleasure. Free flight zones must not be located over populated areas. Traffic areas are only accessible at certain altitudes and landing corridors with the autopilot switched on and make up the majority of the airspace. The airspace is monitored by several satellites simultaneously, creating a three-dimensional flight corridor in which an aircraft moves from take-off to landing. In the medium term, flying cars will take over passenger transport on medium-haul routes and space flights on long-haul routes.

To obtain a flying car licence, you must be observed by a Company Auditing Agency technical auditor performing a flawless 60-minute obstacle flight, including an emergency landing. The licence must be repeated every 10 years.

8.10.2 Space

State Utilities operates an Aerospace Center and participates in international space exploration. The center provides aerospace development and cooperates with the Ministry of Innovation.[103]

Space flights are made possible by new launch facilities at spaceports and offered by State Utilities. Through spaceports, the Ministry of Infrastructure is making space usable.

Sources of income are the transport of persons and cargo as well as the sale of launchers and spaceships as a complete package ready for worldwide use. Another task is the satellite monitoring of space so that flight paths of celestial bodies that could lead to a collision with space ships or the Earth are investigated. This surveillance is carried out in cooperation with the army, which deflects or destroys celestial bodies if necessary.

In the medium term, the spaceports will be used to travel around the Earth in the shortest possible time and spend a short time in space. To this end, the Ministry of Infrastructure is building spaceports all over the world with its Export Division. This reduces the global travel speed and gives more countries the opportunity to use space without polluting it.

In the long term, persons and cargo will be transported from there into space to colonise other planets. To this end, a regular service between Earth and a habitable planet will be instituted.

8.10.2.1 Space junk

The continued deployment of space debris is a crime. Deliberately endangering the Earth to become a prison because its orbit is riddled with billions of small projectiles is considered deliberate killing of future generations. Any service economically performed or imported inland that was

---

103 Ministry of Innovation - 5.4.4 Transport transformation

or is involved in the generation of space debris is subject to a fee. Operators of devices in Earth orbit use it to pay for the research and implementation of the disposal of their space debris. As soon as possible, spacecraft for the disposal of space junk will be developed by the Ministry of Infrastructure and sent into orbit to rid the Earth's orbit of human waste flying around. Since all the parts in orbit have been catalogued, they can be billed to the few possible originators. If other states are not willing to pay, the costs are settled via temporary punitive tariffs.

9 Energy[104]

The Ministry of Infrastructure is responsible for supporting the population with energy. The development of new power sources is done in cooperation with the Ministries of Education and Innovation. The removal of environmental damage in the country is taken over by the originator energy companies.

The Ministry of Infrastructure ensures energy supply and economical energy consumption in the country through laws in energy law and through its institutions. The laws apply to private and state suppliers and consumers of energy. They are designed to ensure that sufficient energy is available to lead today's standard of living and to allow future generations to have at least the same level of energy.

Energy consists of the electricity, heat, transport and industry sectors. Through sector coupling, electricity can be used to produce hydrogen, which can generate heat, power means of transport or be used industrially. Likewise, biogas, like methane, can become electricity or heat. These two gases and exchangeable batteries make it possible to transport electricity without transmission lines. In order to be able to serve all sectors as loss-free and profitably as possible, all sectors are represented in the Ministry of Infrastructure and are supported by the State Utilities and Municipal Utilities Company. This enables the Ministry of Infrastructure, in voting with the Ministries of Health and Innovation, to set overarching energy legislation in all sectors. The State Utilities are responsible for central energy supply and wholesale electricity trading via

104 §205,1,2,3 Energy policy: BV Art. 89

the electricity networks. The Municipal Utilities Companies operate the decentralised energy supply in their municipalities and take care of consumer policy in the energy sector.

Consumers can contact their local Municipal Utilities Company if they have questions or problems with the energy supply. If the questions or problems can only be solved nationwide, the Municipal Utilities Company will pass the case on to the State Utilities Company. In case of doubt, the Council of Ministers must convene. The Minister of Infrastructure then decides with all his deputies or he convenes a committee.[105]

9.1 Security of supply

The supply of energy is diversified so that one source of energy can fail and be replaced by others until it is intact again or an alternative is found. The state energy supply is limited to energy production from naturally occurring and inexhaustible domestic energy sources. These are namely hydropower, gravity, wind power, solar radiation, temperature difference between earth and air or water above and below ground, and the biological decomposition processes of organic material into heat, gas and fertiliser. Surplus energy is stored by pumped-storage power stations and electrolysis for hydrogen production, so that energy is available around the clock all year round and can meet increasing energy consumption.

The safety of the energy plants is checked by the technical auditors of the Company Auditing Agency. Danger from the use of gases, such as methane or hydrogen, is accepted because even in the event of the worst possible accident, only damage that is naturally degradable will occur.

9.2 Economic efficiency

The economic viability of the energy supply results from the properties of renewable energies. Since the moon constantly provides tides, gravity causes water to flow down mountains

---

105 Ministry of State Organisation - 8.6.2 Council of Ministers, 9.6 Committee

into the sea, the sun radiates and interacts with air to provide wind, there are no costs associated with mining, transporting or processing raw materials and no costs for disposal and removal or remediation of mining sites. The more durable and low-maintenance the energy generation plants are, the lower energy can become over time.

Private, corporate and state energy systems support an economic supply. Private individuals occupy buildings that largely support themselves with energy or even generate surpluses. This eliminates the need for a consumer to be supported and creates a producer. Industry can offer more high-quality or cheaper products with more and cheaper electricity. Companies also equip their buildings in the same way as private individuals. In addition, companies can participate in the energy market by operating solar, hydro, wind or biogas power plants. On the one hand, state energy systems consist of energy systems mounted on, in or on state buildings. On the other hand, they consist of solar, hydro, wind or biogas power plants. The state energy plants are operated by the State Utilities and Municipal Utilities Company.

Pumped storage plants and electrolysis have an efficiency of over 70%.[106] Electric motors and generators have an efficiency of over 90%.[107] By comparison, a nuclear power plant, coal-fired power plant or internal combustion engine has a maximum efficiency of 45%.[108] The energy transition will result in savings in raw materials and efficiency, making energy much cheaper and more sustainable in the future.

9.3 Environmental compatibility

An environmentally compatible energy supply is ensured by using renewable energies that interfere as little as possible with the environment and only draw on the energy of the surrounding elements. Although the burning of wood is

---

106https://energie.ch/pumpspeicherkraftwerk/ Stubinitzky, Alexander 2009: Ökoeffizienzanalyse technischer Pfade für die regenerative Bereitstellung von Wasserstoff als Kraftstoff. Accessed: Munich, Technische Univ., Diss., VDI-Verl., Düsseldorf, ISBN 9783183588060.
107https://www.energie-lexikon.info/generator.html
108https://www.energie-lexikon.info/wirkungsgrad.html

considered renewable energy, it is only permitted for private recreational use or in restaurants in a suitable fireplace. Wind turbines erected on land must not cast a cast shadow. Wind turbines therefore have vertical rotor blades and mesh nets to prevent bird damage. Large onshore wind turbines are operated with stunt kites far up in the air. Hydropower plants must not endanger fish and must be shielded accordingly. Solar cells must not contain any non-degradable substances and must not be erected on fertile soil, but on roofs and facades of existing buildings.

9.4 Energy consumption

Economical energy consumption is achieved through insulation, insulation and innovations that have the same or better performance with lower energy consumption. The Ministry of Infrastructure issues laws that set a deadline for mandatory conversion to energy-saving means. Energy consumption is rational if it serves a purpose and provides a benefit. As soon as sufficient renewable energy is available, more and more energy wastage becomes possible. Until then, the Ministry of Infrastructure sets legal limits on consumption or raises the price of energy consumption by facilities, vehicles and devices through a fee. The limits are checked by the technical auditors when awarding the[109] seal of approval. The fees for consumption are determined by the quantity of products sold each year and collected from the originating company as part of the audit by the Company Auditing Agency.

9.5 Energy transition[110]

The energy transition means switching from fossil fuels to renewable energies. Fossil energy sources are uranium, natural gas, oil and coal. Renewable energies use energies that the elements have in them through their interaction.

---

109 Ministry of Labour - 20.7.4.2 Seal of approval
110 §205.6 Energy policy

Fossil energies have led to pollution, which must be eliminated in order to avoid long-term consequences. This is associated with costs that must be borne by the originators. At the same time, the energy transition generates further costs in that renewable energy facilities have to be built and instituted. These economic issues of the energy transition are solved by both energies financing each other. Fossil energies will continue to operate until 120% of the electricity demand can be covered by renewable energies. Those who generate profits with fossil energy sources must transfer them to the Ministry of Infrastructure. In addition, a levy is added to the price of fossil energy sources.

The money is used by the State Utilities, Municipal Utilities Company, private households and companies to connect renewable energies to the electricity network and to build electricity storage facilities. The State Utilities monitor the switchover through measurements and keep energy statistics. As soon as renewable energies provide electricity and gas, the corresponding amount of remaining fossil energy production is discontinued. First, opencast lignite mining is discontinued, then hard coal mining, and finally nuclear energy. How quickly coal mining is stopped depends on the remaining stocks of fuel rods and their remaining useful life. It would make no sense to have uranium fuel for 20 years but still mine coal. Old coal-fired power plants are being converted to use them as energy storage units. They will be used to heat liquid salt, store it and turn it into electricity or district heating when needed.[111] Old nuclear power plants are being converted into liquid salt reactors[112] and operated without radiating waste in the long term.

After 120% of the energy demand can be met, all remaining fossil energy producers will be shut down and disposed of. The electricity price will remain unchanged at a high level, so that once the renewable energies have been financed, the legacy management can be financed. The State Utilities, in cooperation with the originators, will take over the

---

111 https://www.iwr.de/news/kohlekraftwerk-soll-waermespeicher-werden-news35774
112 https://de.wikipedia.org/wiki/Fl%C3%BCssigsalzreaktor

legacy management, namely uranium mine remediation, dismantling and disposal of nuclear facilities, and remediation mining. Mining is rehabilitated in such a way that mines are completely backfilled to prevent uncontrolled subsidence of the earth's surface. Nuclear safety and waste management research is conducted in cooperation with the ministries of education and innovation.

Geothermal energy is a special case. It is no longer operated and boreholes are sealed as best as possible and any cavities created are filled. Since humanity would be extracting large amounts of heat from the Earth, the risks would be similarly devastating as the uncontrolled emission of carbon dioxide into the atmosphere. Extraterrestrial energy such as the moon's gravitational pull and the sun's radiation provide sufficient energy.

The energy transition in mobility is implemented through a deadline to ban the sale of petrochemical fuels. Road users have 5 years to convert their drives.

The energy turnaround is complete when only renewable energies are used and all legacy burdens from the use of fossil energies have been eliminated. On that day, the price of electricity drops to its production costs. Those who cannot support themselves energy self-sufficiently buy energy and have to pay additional line fees for it.

After the energy transition, the people will be able to independently cover their electricity needs from renewable energies within their national borders, without having to purchase energy sources abroad. Industry that needs more electricity is allowed to buy it from abroad, subject to import duties with which the State Utilities increase energy production.

9.5.1 Eternity Task

Closed mines must be cleared of all human waste and pollution. The danger of mine water contaminating drinking water with toxins is just as much a legacy of coal energy as the emission of CO_2 into the atmosphere. These costs must be borne by the originator in order to avoid negative externalities for the population. If these costs have not been priced in, it is a deliberate criminal offence.[113]

In order not to have to pump out mine water forever, disused mines are backfilled with desert sand, which is processed into watertight concrete and glass at the surrounding water-impermeable layers of earth and brought in. Probes are used to control the watertight sealing of the pits and shafts. Should water nevertheless penetrate, the impermeable soil layer must be restored. To do this, it may be necessary to dig tunnels close together with mining robots, which are then filled with waterproof material until watertight levels are created under the earth. The pilot project of such a sealed mine will be operated for 20 years before the process can be transferred to other mines. Pumping out will have to be done for that long. Shutting off the pumps for mine water constitutes the criminal offence of negligent killing of future generations and is punishable by detention for life. The poisoned water must be purified to drinking water quality before it can be used for irrigation or pumped into rivers.

9.6 Energy Directory

The Energy Directory is a digital energy exchange. All suppliers and consumers can trade electricity and heat here. Every energy supplier automatically receives a profile in the Energy Directory after creating a company in the Labour Directory for this gainful activity. Locally close buildings that produce energy can join together in an energy network and thus form a group that can support itself with sufficient energy. Municipal power plants can also set up an energy network whose members supply biomass, for example, and consume the energy generated from it. This trading platform is free of charge and at the same time settles the transmission fees for the use of the state power lines. Consumers report with their profile from the Persons Directory and can purchase electricity or gas.

113Ministry of Justice - 8.6.1 Pollution, 8.7.3 Disposal of contaminated sites, 8.7.4 Energy

9.7 Energy supply[114]

New technologies are constantly being researched for energy supply. The energy turnaround is driven by renewable energies and therefore such construction projects will increase in the short to medium term and will be increasingly researched. The energy supply is organised decentrally and centrally and ensures the security of supply for the people in a nationwide alliance.

9.7.1 Decentralised[115]

The surrounding area of each municipality is different and so are the incomes of the inhabitants. Therefore, decentralised energy supply is the task of the Municipal Utilities Company and the Deputy Minister of Infrastructure is responsible for it. The Municipal Utilities Company itself operates biogas plants and energy storage facilities at suitable locations, as well as wind, river, tidal or solar power plants in parts of the country that are suitable.

The Municipal Utilities Company takes care of collective orders for the citizens of the city who want to equip their buildings with solar panels, wind turbines, zeolite heaters, hydrogen storage heaters or heat exchangers. The demand is determined nationwide. Each citizen of a city is asked to report via a voting in the Energy Directory on the profile page of the local Municipal Utilities Company which form of energy they would like to have, what power the system should have and what they are willing to spend for it as a maximum, with the cost of the product and the installation listed separately. People's Bank offers loans to cover the costs of the conversion. The loans, including interest, are covered by the constant energy price until they are paid off.

114§199.3 State enterprises
115§205.5 Energy policy: BV Art. 89

9.7.1.1 Biogas plant

The Municipal Utilities Company operates biogas plants that are supported by organic waste, natural garden waste and faeces from the sewage treatment plants. Surrounding buildings are connected to the district heating network because heat is released when the waste rots. All of the residents' biological waste is converted into electricity or gas in the same municipality, if possible, depending on how much gas or electricity the residents of the municipality need.

In addition to the biomass reactor, an electrolysis system including a fuel cell and a compressor are also installed in the biogas plant. Via the compressor, methane or hydrogen is pressed into gas tanks located under the biogas plant. These gas tanks serve as the municipality's energy storage. When gas is released, the air pressure is converted into electricity via a turbine. The gas can be converted into electricity in the fuel cell or called up directly via the gas pipeline.

9.7.1.2 River power plant

Paddle wheels are installed at each lock and connected to generators. Water should no longer plunge over an edge into the depths, but flow over a paddle wheel that is as wide as the river. Depending on the level of the river, it is let up or down. Inside the paddle wheel are dynamos that generate electricity. Fish ladders are installed next to the sluices. The paddle wheels can be used as a turning mechanism, for example as a mill wheel, in the event of a disaster and in the vicinity of Barter Economy Zones.

On suitable rivers, the flow velocity is used to generate electricity. To do this, tubes containing turbines and generators are laid at the bottom of a river. The entrances and exits are fitted with grids to prevent fish from swimming in and can be closed automatically. If it is possible to slow down the flow velocity of a river through the current power plants to such an extent that some locks become unnecessary, these locks can be dismantled.

9.7.1.3 Pumped-storage power stations

Municipalities can build water reservoirs at suitable locations to store electricity. For this purpose, water reservoirs are installed in valleys or buildings at two different heights or levels and connected to each other via a pipe including a pump and turbine along with a generator. Surplus electricity is used to fill water through the pump into the upper water reservoir. When electricity is needed, the water is drained from the upper reservoir and falls through the pipe, driving the turbine which generates electricity.

A similar type is used for storage by compressed air. The compressor fills air into a tank with surplus electricity and compresses it under high pressure. When electricity is needed, the air escapes via a turbine that drives a generator.

9.7.1.4 Self-sufficiency buildings

Buildings are ideal for generating renewable energy because they are already connected to the electricity network and usually also to the gas network. In addition, their height enables them to utilise the space three-dimensionally.

The roofs are used to mount solar panels and wind turbines. The solar panels are primarily designed as roof tiles, with solar cells on top to generate electricity and a system of pipes underneath with water to generate hot water. The wind turbines rotate vertically so as not to cast a cast shadow, run in a grid to protect birds and have rough or feathered rotors to avoid noise.

Systems can be installed on the façades and in the boiler room that transport zeolite modules from the boiler room to the façade, where they are dried one after the other in summer and used to power the heating system in winter.

Water tanks are embedded in the floor, which on the one hand absorb rainwater from the roof. On the other hand, they serve the heat exchanger as a temperature differential to the outside or inside temperature of the building. Rainwater from the roof is channelled through rain gutters with built-in generators into water storage tanks between the floors, where it is stored as soon as no further rain falls. It serves as the building's own pumped-storage power station, cooling or heating the building and storing water in case of water shortages. It can then be used for irrigation or domestic water supply. This means that every downpour provides electricity.

Water, liquid salt or hydrogen can be used to store electricity. Water tanks under the roof, between two floors and in the ground create a gradient that is used to operate a small pumped-storage power station. Hydrogen can be produced by an electrolysis plant, stored in a tank and used as a gas for heating or converted into electricity in a fuel cell. The gas can also be fed into the networks via the gas pipeline and sold.

Consumers receive their monthly electricity bill from the Municipal Utilities Company, in which their feed-in to the networks is deducted from the amount of electricity consumed. Accordingly, they receive money or have to pay.

9.7.1.4.1 Order

The above-mentioned products for the generation and storage of energy by buildings, can be ordered via the Procurement Office[116] . In the Energy Directory, a digital ordering platform is offered on each profile of the Municipal Utilities Company. There, orders are collected until a certain order quantity is reached. Through the collective order, a more favourable price is negotiated for the highest-quality product. Negotiations with domestic manufacturers are handled by the Procurement Office. Once a year, the latest and most energy-efficient renewable energy systems for use on buildings are examined, tested and rated. The innovation and technical auditors of the Company Auditing Agency deliver their results annually to the State Utilities, which use them to determine the available range. Only products that offer the best balance between durability, performance and efficiency are selected. The principle applies: "Better invested once than constantly repaired." The lower price comes from the collective orders. Users can choose their suitable product from the list of 30 best products and wait until enough other users do so. Then

116Ministry of Labour - 6 Procurement Office

delivery and payment take place. The installation can be done by craftsmen or by the Construction Team.

9.7.1.4.2 Installation

The Municipal Utilities Company offers a complete solution for energy retrofitting. For this profitable investment in buildings, a personal loan may be applied for from the People's Bank. The interest rate is the base rate that banks also have to pay the Central Bank for money. The loan is paid off by continuing to pay the electricity price of the electricity exchange until the loan is repaid.

First, energy consultants from the Municipal Utilities Company come and, with the help of the computer programmes of the Company Auditing Agency's innovation auditors, check which products the buildings are best suited for and recommend the order. After the order quantity is reached, the delivery and professional installation takes place in the order of the orderers. The Construction Team works through all orders one by one on behalf of the local Municipal Utilities Company, delivers the equipment and then installs it on the building. The Construction Team carries out the construction work as quickly as possible by having as many workers as necessary work on one construction site. If necessary, mobile accommodation is set up for the construction workers and residents or companies in the neighbourhood.

9.7.2 Centralised[117]

The central energy supply is operated by companies and the State Utilities. The Minister of Infrastructure is responsible for security of supply and enacts laws to ensure it. Laws on permissible and eligible energy sources are negotiated with the people in a committee.

The State Utilities adapt the power generation facilities to the geographical and topographical conditions of the areas in the country and use them in the best possible way. The economic

---

117 §199.3 State enterprises, §205.5 Energy policy: BV Art. 89

viability of the plants is ensured through the use of renewable energies, which pay off their initial investment after a period of time. The people decide on the expenditure of state funds for the construction of the plants in the budget vote .[118]
Facilities operated by the State Utilities are pumped storage, marine, solar and wind power plants.

9.7.2.1 Marine power plant

The marine power plants consist of interconnected systems of wind farms in the seas surrounding the continent and underlying wave and tidal power plants. They are installed in the area of the exclusive economic zone at a distance of 20 to 370 kilometres off the coast. Islands that are eroded by tides and currents receive wave, current and tidal power plants off their coast to slow down the water masses. In polluted areas with petrochemical contamination and plastic waste, marine power plants are placed in such a way that buoys are anchored on lines between them to filter water and screen out plastic waste. During maintenance of the marine power plants, the filters on the buoys are changed and the plastic waste is collected.

9.7.2.2 Pumped-storage power plants

The electricity from the marine power plants is used to operate pumped-storage power stations located along the underground network. The underground network connects the north, south, east and west of the country with water pipes and power lines. It uses the natural gradient from the low and high mountains inland to the coast. Upper basins are located in the mountains and catchment basins are located in the valleys and floodplains. Flood basins are used as catchment basins when there is no flood. As the water from the upper basins gradually flows into deeper and deeper catch basins, it passes through turbines in the water pipe of the tunnel tubes, which are connected to generators to produce electricity. The

118Ministry of Finance - 9.5 Budget vote

generators are connected to the adjacent power line via a cable and feed the electricity directly into the networks.

The country should be able to be supported with sufficient electricity solely through the generation of electricity in the sea and the nationwide supply of this electricity through underground cabling as well as through the nationwide decentralised storage points.

Any excess capacity in the electricity network, from wind or solar power, is fed into the pumped storage facilities. This pumps the water back from the catchment basins into the upper basins.

In addition to pumped storage plants for water, surplus electricity is used for electrolysis to produce hydrogen. The hydrogen is pumped into tanks under high pressure and stored. When it is released, turbines are driven with the compressed air to generate electricity via a generator. The hydrogen can then be used as fuel to generate heat or to produce electricity in a fuel cell. The efficiency of a fuel cell is currently between 40 and 80 percent.[119]

9.7.2.3 Solar power plant

Solar power plants are installations with mirrors, burning glasses and combustion chambers in which fluid is heated and expands, creating pressure that is passed through a turbine, which then drives a generator that produces electricity. These power plants are placed in locations that receive as many hours of sunlight as possible during the year. Solar cells that are placed in fields and thus displace farmland are not permitted.

9.7.2.4 Solar balloon

Helium balloons are released on an electric cable to a height of about 200 metres. The higher they rise, the more they unfold. Solar cells are printed on foil on the outer shell, which generate electricity and feed it into the networks via the power cable.

---

[119] https://www.energieagentur.nrw/brennstoffzelle/brennstoffzelle-wasserstoff-elektromobilitaet/wirkungsgrad

Solar balloons are a way to provide electricity for short periods anywhere. State Utilities offer trucks with solar balloons and batteries for hire. In suitable places, solar balloons can also be anchored permanently and operate electrolysis as storage.

9.7.2.5 Sail power plant

Sail power plants are generators mounted in a ring that lies horizontally on the ground and has a diameter of about 200 metres. The generators are attached to the rigid side of the ring and stunt kites are attached to the rotating side of the ring. The stunt kites act like sails and are managed automatically to stay in the wind as best as possible. They fly at an altitude of about 500 metres, but adjust their flight altitude to the height where the most wind is blowing at the moment. To do this, they carry sensors on the sail and the strings that measure the wind direction and strength. The faster the ring turns, the more electricity is generated. The State Utilities place sail power plants at suitable locations, for example as a kind of crown for mountain peaks or in coastal regions. In the middle of the ring at the bottom is a cylindrical pumped-storage power stations. A cylinder about 50 metres in diameter is milled into the ground and its base is installed at a depth of about 50 metres. The soil inside it remains unchanged. Water can be pumped under the cylinder so that it rises up to about 30 metres from the ground. When electricity is needed, the cylinder is lowered and pushes the water through generators that produce electricity.[120] The State Utilities also have trucks that carry a stunt kite and batteries. Via a rotating crane, the stunt kite is first set in motion to gain altitude until it is at a flying height with sufficient wind.[121]

---

120 https://heindl-energy.com/
121 https://enerkite.de/

9.7.2.6 Wind power plant

Existing electricity pylons located in places with sufficient wind are given a vertical wind turbine in their centre. The wind can come from all sides and the windmill immediately starts to turn. At the top is a wind meter that detects the wind strength and switches on more generators to produce more electricity when the wind is strong. At the bottom of the wind turbine are several generators one below the other.

As soon as the underground cabling is completed, overland power lines can be dismantled. The freed-up electricity pylons are retrofitted with wind turbines and erected in places with sufficient wind. The State Utilities provide these power poles to the Municipal Utilities Company or sell them to companies or private individuals for power generation.

9.8 Development of new power sources[122]

The Ministry of Infrastructure is working with the ministries of innovation and education to develop new sources of electricity. Through research projects of the Ministry of Innovation, educational institutions, companies and Innovation Labs are conducting research on how to further harness renewable energy.[123] The major direction of research is in biotechnology. Future energy sources will come from a comprehensive understanding of photosynthesis, the use of adenosine triphosphate (ATP) and how animals can generate heat and light from it. In the long term, chloroplasts in muscle cells are capable of driving generators that produce electricity. As a closed circuit, these muscles only need water, minerals and sun to do work. This is available in large quantities in the world's oceans.

The Ministry of Infrastructure has the task of supporting the research institutions in testing new energy generators. Test facilities are built that operate under real conditions and are continuously examined by the researchers. State researchers are allowed to request the Ministry of Infrastructure to carry out

122§205.3 Energy policy: BV Art. 89
123Ministry of Innovation - 5.4.3 Energy Transition, 6.5 Mobile Innovation Labs

studies in their plants and power stations, the measurement and test results of which are sent to the researchers and evaluated by them. The aim is to increase efficiency, conduct basic research and test further developments.

9.8.1 Tesla laboratory

The researcher Nicola Tesla had groundbreaking inventions and a comprehensive understanding of electricity. In the Tesla Labs, all the knowledge surrounding Nicola Tesla, including all the "conspiracy theories", is collected in order to continue his research. The Tesla Lab is set up at a suitable university and can contribute individual experiments and experimental set-ups to the Innovation Labs or create electrical kits for teaching at educational institutions.

9.8.2 Lightning power plant

In the mountains and in regions where thunderstorms occur frequently, power lines are stretched to serve as lightning conductors and conduct the electricity into a transformer. The transformer still has to be developed to be able to catch and split the high voltage of a lightning strike. This is to be used to operate pumped storage or electrolysis or to charge batteries.

10 Switching to the new system

For the changeover, all affected ministries must hand over their departments and reunite them with the Ministry of Infrastructure. First of all, all construction activities will be directed towards the construction of the Social Villages, which will also accommodate asylum seekers at the beginning. Asylum seekers work with the Construction Team after the Social Villages have been built. Their mobile shelters made of containers serve as living quarters for the Construction Team. The production sites for the infrastructurators, containers and building materials are built in structurally weak regions where infrastructures of past industries are used.

10.1 Emissions trading

Emissions trading requires companies that release harmful emissions to buy So-called pollution allowances for the emissions. The purchase price of a certificate depends on how much it currently costs to neutralise the emission. The introduction of environmentally friendly innovations can lower the purchase price.

Emissions trading between the companies is discontinued and switched to trade with the State Utilities until the profit levies and price surcharges from the "Energy Turnaround" chapter above take effect. Every company that pollutes the environment through production or goods and services must buy suitable certificates for this from the State Utilities. The price increases by 10% every year. Certificates vary in price depending on the type of pollution and are only valid for a certain amount of pollution. For each additional quantity, certificates must be bought again. It is only permissible to buy certificates for the period of one year. The certificates must be paid for at the end of the calendar year. The economic auditors of the Company Auditing Agency check the quantities and settle the corresponding number of certificates and transfer the revenues to the State Utilities.

The prices are based on the amount of damage, but the revenues are used by the State Utilities to build renewable energy plants. The damage is repaired with the revenues from the renewable energy plants.

The innovation auditors of the Company Auditing Agency examine the affected companies and propose innovations to them that can reduce the purchase price of the certificates. To this end, companies are first examined to see whether production can be converted. The innovation auditors advise whether there are environmentally friendly production methods, materials, raw materials, recycling or disposal options or whether filter systems can be installed, such as farms made of algae and microbe batteries that can metabolise environmentally harmful gases directly on site.

10.2 Changeover in the construction industry

The Building Offices network and the Ministry of Infrastructure establishes the National Building Office. The Building Office ensures the establishment of the construction team in cooperation with the Building Yards.

Much of the military's equipment in construction machinery and personnel is placed under the Ministry of Infrastructure in peacetime. Voluntary former soldiers and operational equipment without weapons or ammunition are deployed in the Construction Team on construction sites.

State agencies for geosciences and raw materials, which will be transformed into the Institute for Geosciences and Raw Materials, will be responsible for the exploration of raw materials for construction and suitable building sites. State agencies for the digitalisation of planning, construction and operation will be integrated into and build up the Infrastructure Directory.

10.2.1 Housing construction

No families with children should have to live in apartment blocks, i.e. prefabricated buildings with more than 6 storeys. They are entitled to state-owned flats and houses and can buy them on a lease-purchase basis. Excluded from this are all areas designated to become Social Villages with Planned Economy. These include former barracks and brownfield sites. The proceeds from the Lease-purchase are used for the expansion of the Social Villages by the Construction Team.

All apartment blocks are sold or rented to unmated persons or childless couples, converted, deconstructed or demolished. The conversion can turn each floor into a flat, convert floors into gardens with aquariums, or the buildings become common property and all the castles are removed. The use must be voted on democratically among all the guests. The apartment blocks are deconstructed by retaining only the lower two floors and converting them into detached or semi-detached houses.

Social housing was characterised by high-rise buildings. In

the future, a maximum of three storeys will be built on top of each other and each residential house will have a garden. The housebuilding programme is first implemented for the Asylum Villages of the asylum seekers and becomes a pilot project for the Ministry of Infrastructure. The Construction Team here consists mostly of asylum seekers who build their own village to live in until deportation.[124] These houses are then transferred to nationals by Lease-purchase.

10.3 Switching to the new networks

First, all Municipal Utilities Companies in cities and municipalities where they have been privatised are to be nationalised. The state agencies for operating telecommunication and radio networks will be transformed into the network agency. The money from the digitalisation funds will be spent to build the intranet. Coal mining will be stopped as soon as possible. Coal miners from all collieries will be offered a transfer to the Construction Team and will be employed there mainly to build tunnels.

10.4 Changeover in transport

First, the roads will be approved from Truck traffic and the transport of goods will be shifted to the railways. Vehicle tax will be abolished and tolls introduced. The automobile industry will be forced to deliver an exchangeable battery with standardised dimensions and connections in their electric cars. Filling stations will be forced to be able to exchange and charge exchangeable batteries. In and around large cities, a sufficient number of roads for motor vehicle traffic must be converted into cycle lanes. The blue sticker applies to hybrid vehicles with internal combustion engines and the purple sticker applies to vehicles powered by electricity or hydrogen. From 2025, cities may only be entered with a blue sticker, and from 2030 only with a violet sticker. From 2035, there will be a general ban on driving with internal combustion engines.

---

124 Ministry of Integration - 8.6.1 Asylum Village

The automotive industry will be forced to convert its vehicles with internal combustion engines that are less than 15 years old to electric or hydrogen drive. Only the drive system will be changed, because a new production to replace all vehicles would be too expensive and environmentally harmful. For example, there are wheel hub motors and replaceable batteries where the internal combustion engine was.

The Traffic Office brings together all matters and regularisations involving motor vehicles, railways, shipping, aviation, space, freight transport and tolls on roads. A department in the Traffic Office takes over the Council representation at the International Civil Aviation Organization (ICAO). The technical auditors of the Company Auditing Agency carry out the safety inspections of the means of transport and routes.

The State Utilities Company and the Municipal Utilities Company take on all the tasks of child support for roads, railways, air traffic control, airways and waterways. The assets from rail transport flow into the switch of freight transport to rail and the development of the Maglev network.

10.5 Conversion of the old ministries

For the conversion of the old ministries, all departments and units of the old ministries that are changing to this ministry are identified. The organigrams are used to determine whether an entire department and all its units are changing or only individual units. All unsuitable departments and units are dropped. The existing staff adapts its tasks to the new requirements.

Contact form

Dear reader

If you would like to make what you have read come true, in whole or in part, together with other like-minded people, I offer you several possibilities with this contact form. Fill it out, tear out the page and send it by post to:
Andreas Seidl, P.O. Box 1206, 63488 Seligenstadt / Germany

Or send the details to:
Phone: 0049 1522 818 2243 (whatsapp, telegram, signal)
Email: andreas.seidl2022@web.de

Please mark with a cross:
O   I want to found a dynamic People's Party.
O   I want to donate money for implementation.
O   I want contacts with like-minded people in my area.

Forename: ____

Surname: ____

Please fill in only the contact option through which a reply should be made.

Street, house no.:   ____________________________________

Postcode, city, country: ____________________________________

Phone: ____________________________________

Email address: ____________________________________